ENERGY SCIENCE, ENGINEERING AND TECHNOLOGY

INCORPORATING SOLAR TECHNOLOGIES IN THE UTILITY RESOURCE PLANNING PROCESS

ENERGY SCIENCE, ENGINEERING AND TECHNOLOGY

Additional books in this series can be found on Nova's website
under the Series tab.

Additional e-books in this series can be found on Nova's website
under the e-book tab.

ENERGY SCIENCE, ENGINEERING AND TECHNOLOGY

INCORPORATING SOLAR TECHNOLOGIES IN THE UTILITY RESOURCE PLANNING PROCESS

SANDRA M. GROSSO
EDITOR

nova publishers
New York

Library of Congress Cataloging-in-Publication Data

ISBN: 978-1-63117-009-6

Published by Nova Science Publishers, Inc. † New York

CONTENTS

PREFACE

Today's utility planners have a different market and economic context than their predecessors, including planning for the growth of renewable energy. State and federal support policies, solar photovoltaic (PV) price declines, and the introduction of new business models for solar PV "ownership" are leading to increasing interest in solar technologies (especially PV); however, solar introduces myriad new variables into the utility resource planning decision. This book focuses on the treatment of solar generation in electric utility resource planning and provides and evaluation of solar valuation methods used in utility planning and procurement processes.

Chapter 1 – Today's utility planners have a different market and economic context than their predecessors, including planning for the growth of renewable energy. State and federal support policies, solar photovoltaic (PV) price declines, and the introduction of new business models for solar PV "ownership" are leading to increasing interest in solar technologies[1] (especially PV); however, solar introduces myriad new variables into the utility resource planning decision. Most, but not all, utility planners have less experience analyzing solar than conventional generation as part of capacity planning, portfolio evaluation, and resource procurement decisions. To begin to build this knowledge, utility staff expressed interest in one effort: utility exchanges regarding data, methods, challenges, and solutions for incorporating solar in the planning process. Through interviews and a questionnaire, this report aims to begin this exchange of information and capture utility-provided information about: 1) how various utilities approach long-range resource planning; 2) methods and tools utilities use to conduct resource planning; and, 3) how solar technologies are considered in the resource planning process.

The National Renewable Energy Laboratory (NREL) and the Solar Electric Power Association (SEPA) worked together to engage utilities directly to research this topic and author this report. An advisory council of electric industry experts guided the methodology used and provided feedback on draft analysis. The main sources of information captured in this report were predominantly from utilities. The authors conducted interviews with electric sector representatives from 13 entities (9 of which were utilities) and developed a utility questionnaire that secured more specific modeling data sources and methodologies from 28 utilities in 22 states. Key questions from the questionnaire and a summary of utility responses are included in Appendix B.

Chapter 2 – As renewable technologies mature, recognizing and evaluating their economic value will become increasingly important for justifying their expanded use. This report reviews a recent sample of U.S. load-serving entity (LSE) planning studies and procurement processes to identify how current practices reflect the drivers of solar's economic value. In particular, The authors analyze the LSEs' treatment of the capacity value, energy value, and integration costs of solar energy; the LSEs' treatment of other factors including the risk reduction value of solar, impacts to the transmission and distribution system, and options that might mitigate solar variability and uncertainty; the methods LSEs use to design candidate portfolios of resources for evaluation within the studies; and the approaches LSEs use to evaluate the economic attractiveness of bids during procurement.

We found that many LSEs have a framework to capture and evaluate solar's value, but approaches varied widely: only a few studies appeared to complement the framework with detailed analysis of key factors such as capacity credits, integration costs, and tradeoffs between distributed and utility-scale photovoltaics. Full evaluation of the costs and benefits of solar requires that a variety of solar options are included in a diverse set of candidate portfolios. The design of candidate portfolios evaluated in the studies, particularly regarding the methods used to rank potential resource options, can be improved. The authors found that studies account for the capacity value of solar, though capacity credit estimates with increasing penetration can be improved. Furthermore, while most LSEs have the right approach and tools to evaluate the energy value of solar, improvements remain possible, particularly in estimating solar integration costs used to adjust energy value. Transmission and distribution benefits, or costs, related to solar are rarely included in studies. Similarly, few LSE planning studies can reflect the full range of potential benefits from adding thermal storage and/or natural gas augmentation to concentrating solar power plants. Finally, the level of detail provided in requests for proposals used in procurement is not always sufficient for bidders to identify the most valuable technology or configurations to the LSE.

In: Incorporating Solar Technologies in the Utility Resource ... ISBN: 978-1-63117-009-6
Editor: Sandra M. Grosso

Chapter 1

TREATMENT OF SOLAR GENERATION IN ELECTRIC UTILITY RESOURCE PLANNING*

John Sterling, Joyce McLaren, Mike Taylor and Karlynn Cory

LIST OF ACRONYMS AND ABBREVIATIONS

AC	alternating current
APS	Arizona Public Service
CA IOUs	California investor-owned utilities
CEM	capacity expansion model
CPV	concentrating photovoltaics
CSP	concentrating solar power
DC	direct current
DG	distributed generation
DOE	Department of Energy
Duke	Duke Energy
EE	energy efficiency
EERS	energy efficiency resource standard
ELCC	Effective Load-Carrying Capability
GHG	greenhouse gas
Idaho	Idaho Power
IID	Imperial Irrigation District
IRP	integrated resource plan
IOU	investor-owned utility
ITC	Investment Tax Credit
LADWP	Los Angeles Department of Water and Power
LBNL	Lawrence Berkeley National Laboratory
LOLP	Loss of Load Probability

* This is an edited, reformatted and augmented version of NREL is a national laboratory of the U.S. Department of Energy Office of Energy Efficiency & Renewable Energy, NREL/TP-6A20-60047, dated October, 2013.

MW	megawatt
MWh	megawatt-hour
NEM	net energy metering
NREL	National Renewable Energy Laboratory
NVE	NV Energy (Nevada)
PCM	production cost model
PGE	Portland General Electric
PNM	Public Service of New Mexico
PPA	purchased power agreement
PSCo	Public Service of Colorado
PV	photovoltaic
PVRR	Present Value Revenue Requirements
RE	renewable energy (or electricity)
RFP	request for proposals
RPS	renewable portfolio standard
SCANA	An electric and gas utility in SC and NC
SEPA	Solar Electric Power Association
SREC	solar renewable energy credit
T&D	transmission and distribution
TEP	Tucson Electric Power
TMY	typical meteorological year
Tri-State	Tri-State Generation and Transmission Association

EXECUTIVE SUMMARY

Today's utility planners have a different market and economic context than their predecessors, including planning for the growth of renewable energy. State and federal support policies, solar photovoltaic (PV) price declines, and the introduction of new business models for solar PV "ownership" are leading to increasing interest in solar technologies[1] (especially PV); however, solar introduces myriad new variables into the utility resource planning decision. Most, but not all, utility planners have less experience analyzing solar than conventional generation as part of capacity planning, portfolio evaluation, and resource procurement decisions. To begin to build this knowledge, utility staff expressed interest in one effort: utility exchanges regarding data, methods, challenges, and solutions for incorporating solar in the planning process. Through interviews and a questionnaire, this report aims to begin this exchange of information and capture utility-provided information about: 1) how various utilities approach long-range resource planning; 2) methods and tools utilities use to conduct resource planning; and, 3) how solar technologies are considered in the resource planning process.

The National Renewable Energy Laboratory (NREL) and the Solar Electric Power Association (SEPA) worked together to engage utilities directly to research this topic and author this report. An advisory council of electric industry experts guided the methodology used and provided feedback on draft analysis. The main sources of information captured in this report were predominantly from utilities. The authors conducted interviews with electric sector representatives from 13 entities (9 of which were utilities) and developed a utility questionnaire that secured more specific modeling data sources and methodologies from 28 utilities in 22 states. Key questions from the questionnaire and a summary of utility responses are included in Appendix B.

Resource Planning Background

Integrated resource plan (IRP) procedures and long-term planning refer to the processes utilities[2] take to evaluate a wide range of potential supply- and demand-side resource options to meet energy (MWh) requirements and peak demand (MW), plus a reserve margin. Utilities look for the best set of future decisions to reliably serve customers and meet regulatory requirements over a long-term period, usually approximately 20 years. This process is traditionally designed to maintain reliability and meet load at the lowest reasonable cost (i.e., least-cost). In recent decades, IRP processes have included other factors, including renewable energy mandates. Additionally, some utilities have recognized that generation diversity provides direct benefits and utilities have started to place increased emphasis on lowering risk and uncertainty of future regulations and fuel prices, on "affordable" costs, or other non-cost metrics (unless prevented by state law or statute). Valuing diversity and considering the risk of the status quo is one way utilities are starting to incorporate non-cost metrics in decision-making. Some utilities note that an exclusive focus on least-cost may not yield the optimal portfolio of resources.

Through interviews and a questionnaire, the authors gathered information on utility supply planning. Utilities were asked to provide their resource planning process details, key assumptions (e.g. whether DG is represented as supply or negative load), modeling methodology (e.g. type of risk analytics and candidate portfolio development), and capacity expansion and production simulation model software. Utilities performing the most detailed long-term planning[3] include the following steps (and within steps there are many variations):

1. **Evaluate State Policies and Mandates**. Evaluate state laws and regulations that influence future electricity demand and supply (e.g. efficiency and renewable energy mandates), because they impact the amount and timing of new supply procurement needs and the characteristics of demand increases through the year and over time.
2. **Review Existing Generation Fleet**. Consider existing baseload, intermediate, peaking, and variable renewable generation that is owned or under a power purchase agreement (PPA).
3. **Forecast Load**. Study each customer class and forecast current and future energy and demand requirements. Roll up the forecasts to determine peak demand periods on a daily, monthly, seasonal and annual basis. Attempt to anticipate future disruptive changes that could increase demand (e.g., plasma televisions and electric vehicles), decrease demand (e.g., energy efficiency program success), or moderate demand (e.g., smart meters, load shifting/shedding).

4. **Plan Capacity Expansion**. Analyze options for meeting long-term demand using either capacity expansion modeling or engineering judgment. Popular commercial models include Strategist, System Optimizer (previously known as Capacity Expansion), AURORAxmp and EGEAS, although a few utility interviewees use in-house models. These models can include one or more specific constraints, such as: specific resource limits, minimum renewable requirements, transmission plans and capabilities, resource restrictions, and forced plant additions or retirements. Generally speaking, the goal of the capacity expansion process is to minimize future costs, given the constraints in question. While not always available in commercial models, there are a wide range of supply- and demand-side technologies and a variety of configurations possible. Utilities interviewed wish more technology options were available in commercial models. Most utilities consider scenarios /plausible futures that need to be tested (e.g., forced plant retirements, carbon tax, or high natural gas prices) to ensure a robust process.
5. **Production Cost Modeling**. Simulate hourly dispatch over the entire planning horizon and perform complex sensitivity analyses by: 1) using Monte Carlo simulation to randomly change multiple key input variables hundreds or even thousands of times; [4] 2) varying single variables to examine specific impacts of major assumptions on a portfolio; or 3) a combination of both. Variables that can be modified include fossil fuel prices, wholesale market prices, load, environmental costs, renewable energy levels, hydro availability, energy efficiency, and incentive availability.
6. **Select Portfolio.** Select a portfolio, in which key metrics for each mix of future resources are tracked for comparison purposes. Metrics include revenue requirements, capital expenditures, emissions, fuel diversity, water usage, and average system cost. Future risk (including generator diversity and potential regulations) is a more recent metric used by some utilities to incorporate potential future risks into today's decision- making. To select a resource portfolio that is both low cost and resilient to upward price risk, utilities can focus on the expected Present Value Revenue Requirements (PVRR) as well as the potential price risk (or risk tail, represented by the upper 5% of PVRRs or some other value).

Inclusion of Solar in Resource Planning Processes

Through interviews and a questionnaire, the authors gathered information on solar project representation in long-term utility plans (project size, capacity value and integration cost adder).

Interviewees gathered solar PV profile and cost information from a variety of sources. Solar generation profiles were either taken from existing operational solar plants, or from external sources like PVsyst or NREL's PVWatts program, which provide a typical year of data for a limited number of specific locations. While there is interest in the combination of storage and PV, storage is not modeled in long-term planning by the majority of utility interviewees, because utilities feel they need more credible data and analysis before storage can be successfully incorporated.

Solar cost information was less consistent – and there were big questions about the future, especially for PV technologies. Some utilities interviewed believed that U.S. Department of Energy SunShot goals for PV may be achieved by 2020; others said costs will continue to go down but at a slower pace, and still others said that solar costs have reached their minimum and will flatten out for the foreseeable future. These solar PV cost assumptions are reflected in their solar cost assumptions used in their long-term planning. Notably, utilities indicated they would like more credible estimates of future solar PV prices, because predicting future solar PV prices is considered a key uncertainty, especially over a multi-decade planning horizon. While prices have come down rapidly, questions remain about how long the prices will continue to decline and when they will bottom out (or if they already have).

One key question discussed at length with interviewees was the capacity value utilities attribute to the addition of solar capacity on their system. In essence, a solar generator's capacity value is the percentage of its nameplate capacity that is anticipated to be reliably available to meet daily and seasonal peak demand, and is a region-specific metric.[5] Fully dispatchable conventional generation can be called upon to meet load as it fluctuates throughout the day and year. In contrast, the most commonly deployed solar technology, fixed-axis solar PV, generates only when the sun is available. Maximum energy production is mid-day; as the sun sets the production capability can diminish rapidly, depending on the panel orientation (westward panels can moderate this impact, although overall production is likely reduced). Most utilities peak later in the afternoon (e.g., between 4-6 p.m. in the summer months), so solar energy may not correlate perfectly with a utility's need, depending on their specific situation (e.g., latitude, insolation). Different solar technology configurations can moderate this potential for rapid impact through the addition of storage, or PV tracking (one or multiple axes).

Figure ES 1 shows the ranges of capacity values attributed to a variety of solar configurations, based on questionnaire responses. 60–76% of respondents assigned some capacity value to the first three solar PV configurations, ranging from 0–60% for fixed PV and up to 80% with tracking. Conversely, only 26–44% of utility respondents assigned any capacity value to CSP (with or without storage), PV with battery storage or concentrating photovoltaics (CPV); 56– 74% used a 0% capacity value.[6]

A few other considerations for including solar in resource planning were briefly discussed during interviews, including solar integration costs (the costs to manage solar variability with other generating resources, typically $2-11/MWh – as indicated by utility IRPs investigated by LBNL (Mills and Wiser 2012)), whether to treat customer-sited distributed generation as net load (typical) or as a resource (just starting), and ways to link long-term resource planning with power procurement through request for proposal processes.

A number of key challenges of incorporating solar into supply planning analysis were raised during the utility interviews. Most of these utilities have not experienced notable levels of solar PV, CSP or CPV penetration, and have not always modeled solar in their IRP studies. Most utilities expressed an interest in incorporating and refining distributed- and utility-scale solar PV in their modeling (since they anticipate more PV coming onto their system in the next decade), although some utilities noted that CSP and CPV are not appropriate technologies for their service territories. Several utilities mentioned that they participated in this project specifically to learn from their peers about best practices.

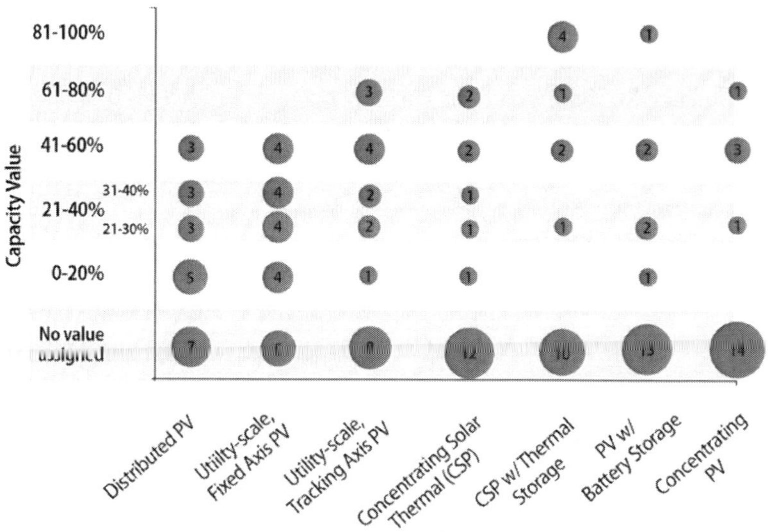

Note: Numbers in circles represent the number of utility responses.

Figure ES 1. Range of solar capacity values used in planning analysis, by technology, provided by utility questionnaire respondents.

Table ES 1. Benefits and Challenges of Solar, Based on Utility Interviews

Benefits of Solar	Challenges of Solar
• Meet renewable standard requirements • Fuel diversification • Cost stability • Geographic dispersal benefits and modularity • Partial correlation with peak demand • Mitigation of environmental compliance risks • Avoid line losses (typically DG only)	• Variableand uncertain output • Ramping issues • Economics • Lack of current capacity need • Cross-subsidization concerns (DG) • Reduced capacity benefit over time with increasing solar penetration

Improving Solar Within Utility Supply Planning

Utility interviewees identified a number of benefits and challenges associated with solar technology, listed in Table ES 1 and described in more detail in the report. And with declining technology costs and increasing customer interest, the participants stated that they are likely to include solar PV as a resource in their forthcoming plans.

With these benefits and challenges in mind, utilities can more accurately incorporate solar generation into their long-term planning processes. The following list highlights some of the leading, utility-identified best-practices and analysis needs discussed in Sections 4 and 5 of this report:

1. ***Analyze and assign appropriate capacity values to solar resources*** – While not always available at 100% of potential output when a utility's demand peaks, solar generation provides some ability to meet that peak. Utilities can carry out capacity valuation analyses for solar resources that are specific to their system and region (PV

with or without tracking and with or without storage, CSP with or without storage and CPV). For PV, differentiation between fixed tilt and tracking systems will clarify results. Further, utilities can update all solar values on a routine basis as penetrations increase (which may result in declining capacity values over time). Consider the benefits of geographic diversity and be sure to include integration cost add-ons, if needed.

2. ***Analyze solar individually, to get more accurate aggregate results*** – Individual solar assets are subject to output variability relative to the specific conditions at their sites; however, when solar generation is aggregated across multiple systems in various sites this variability becomes substantially reduced. For long-term planning purposes, consider aggregating individual solar profiles to create a more accurate aggregated geographic dispersion of specific systems.
3. ***Improve modeling assumptions and methods*** – If not modeled accurately, the inclusion of solar can yield erroneous results. Based on analysis experience to date, results can be skewed when solar is added to *previously-optimized* generation portfolio, decreased marginal capacity credit is not accounting for with higher solar penetrations, neighboring utility activities are unaccounted for, or accurate thermal generator operating limits are not appropriately captured.
4. ***Pursue sub-hourly sensitivities*** – The vast majority of production cost modeling work undertaken today is done based on an hourly dispatch over the planning horizon. Attempting to run those same models on a more granular level (such as 15-minute dispatch) would prove both time-consuming and computer-intensive. As an alternative, utilities can consider modeling sample periods (such as one week a season for each planning year) on a more granular sub-hourly basis. This approach could provide insight into the integration issues caused by variable resources like solar, show the impacts to generation ramping, dispatch, and system operations, and lead to refinements in the amount of flexible resources or energy storage included in future portfolios of resources.
5. ***Evaluate whether to treat distributed generation as a resource*** – The majority of utilities today treat distributed generation (DG) resources as a net load impact. This approach, while acceptable, in essence embeds distributed solar penetration variability within the utility's load forecast. Utility load is one of the key drivers of long-term supply planning, so combining DG and actual load forecast may make it harder to determine the impact of either alone. Alternatively, it might make sense to analyze solar DG as a resource that can be dynamically added in capacity expansion models. Utilities can then test how changes to key input variables impact the cost viability of distributed solar resources, and begin to optimize the amount of DG they want to target in the future.
6. ***Utility-identified analysis needs*** – Finally, utilities identified a number of key analysis needs that would better inform their supply planning, including:
   - Credible PV price and performance data (preferable from other utilities or another third-party; not from industry)
   - Analysis of how to incorporate geographically diversified resources into modeling
     - Example: Instead of one 50 MW PV plant; model ten 5 MW PV across territory

- Analysis of the potential relationship between energy storage and PV
- Easier ways to predict impacts of increased PV penetration
- Better risk/uncertainty analysis methods (beyond scenario planning)
- Improved commercial models that include:
    - Currently unavailable technologies (e.g. solar PV, CPV, CSP, wind, etc.)
    - Intra-hour sensitivity capability
    - Improved performance (e.g. runtime when running sub-hourly)
- Capture distribution system impacts of DG and other technologies/activities in long- term plans (broader issue than solar)
- Clarity about when to include DG in supply modeling
    - What is the tipping point that will impact results?
    - What is net-energy metered DG's impact on revenue requirements?[7]
- Supply additions during periods of low/no load growth.

1. Introduction

1.1. The Task for Utility Resource Planners

When large coal facilities were constructed 50 years ago, utility planners had never heard of production cost models or integrated resource plans. Climate change was still an unfamiliar term, and utilities were tasked with meeting a steady six-percent load growth. Committing to large coal-fired generating stations, despite the risk and investments they required, was likely viewed as a necessity to provide baseload generation, rather than one option of many.

Today's utility planners are presented with a very different landscape: near-zero load growth, the near-term retirement of large baseload facilities, fuel price uncertainties, a variety of evolving generation and grid technologies, state and federal efficiency and renewable support policies, and a host of environmental considerations. Planners are asked to balance a wide variety of factors relating to reliability, cost, risk, environmental protection, and equity. They have an aging grid, new customer demands, and a host of stakeholders to consider. Utility resource planning has evolved into a resource-intensive process that requires significant expertise, substantial time, and sophisticated tools (Wilson and Peterson 2011). Even so, there are many benefits to undertaking a thorough planning process to determine the lowest practical costs of reliably meeting expected future electricity demand. For this reason, at least 27 states now require utilities to produce formal IRPs or similar plans (Wilson and Biewald 2013).[8]

1.2. Purpose of this Report

One of the relatively new elements that resource planners are challenged to consider is solar energy technology.[9] The goal of this report is to review current utility practices in long-term supply planning, and to understand how solar technologies are represented and modeled.

Federal and state policy developments, solar PV price declines, and the introduction of new solar PV "ownership" business models are stimulating increasing interest in solar technologies, at every scale. However, solar introduces a host of technology types, a wide variety of system sizes (from residential to utility-scale), variability in the generation profile, and rapidly changing costs. In addition, the established rules, regulations, and laws that govern utility supply procurement were developed during a time when large-scale, centrally-located, conventionally-fueled generation was the only real option considered. Now that generation options and different configurations are increasing, the policies may need to be adapted to new technological, operational, and market needs, such as distributed solar generation.

Most, but not all, utility planners have less experience analyzing solar than conventional generation as part of capacity planning, portfolio evaluation, and resource procurement decisions. In the last 5-10 years, some utilities have developed tools and methods to model solar PV, if not other solar technologies. However, aside from publicly-available IRPs, there have been few resources through which utility planners can learn from each other regarding methods, challenges, and solutions about how to consider solar in planning process.[10] This report aims to begin this exchange of information.

In sum, this report aims to build understanding of:

- Time frames and methods for utility long-range resource planning
- Tools and assumptions to conduct utility resource planning
- How solar technologies are considered in the resource planning process.

1.3. Key Audiences

The primary audience for this report is utility resource planning departments. The report's aim is to share information about how others in the industry incorporate solar generation – both utility-scale and customer-sited – into their resource planning processes. A secondary target audience is the U.S. Department of Energy, who is interested in understanding how solar electricity is represented in the market and whether there are analysis needs the Department can address.

1.4. Related Research to Date

The research for this report builds upon work completed in 2012 by Lawrence Berkeley National Laboratory (LBNL), as reported in "An Evaluation of Solar Valuation Methods Used in Utility Planning and Procurement Processes" (Mills and Wiser 2012). For that effort, LBNL conducted a literature review of a sample of utility planning studies and procurement processes. Their aim was to understand the methods utilities use to design resource portfolios for consideration, as well as how the utilities evaluate the various bids during the procurement process. They looked at how the utilities approached a variety of factors – including capacity value, energy value and integration costs, risk reduction, grid integration, and operability issues – as they relate to solar.

The LBNL team found that utility approaches vary widely. While many utilities have a method to evaluate the value of solar, only a few conducted detailed analyses that considered key factors. Rarely were the tradeoffs between various solar options or the full range of benefits and costs to the grid considered. The authors noted that improvements could be made to current utility methods in many areas, including the ranking of resource options, estimating capacity, evaluating the energy value of solar, estimating integration costs, and considering storage.

This report expands on the LBNL report. While the LBNL report used information found in utility IRPs and looked only at methodology, this report relies on interviews and a questionnaire to investigate the planning process and assumptions more thoroughly.

1.5. Research Methodology

1.5.1. Research Team

NREL managed the project, directed research, and led analysis of results. Together, NREL and SEPA conducted electric sector interviews, delivered a utility questionnaire, and analyzed results. NREL and SEPA worked together to author this report.

1.5.2. Advisory Council

Prior to beginning the research for this report, a small advisory council was formed. The council helped focus the project's goals, inform its design and review the questions that were used in interviews with select utilities and in a written questionnaire. The council consisted of 6 members representing two utilities, two national labs (NREL and LBNL), and two stakeholder groups (National Regulatory Research Institute and Western Resource Advocates).

1.5.3. Utility and Stakeholder Participation

Much of the data for this report was obtained through interviews with utility planners and non-utilities involved with the resource planning process, including software providers and consultants. Most, but not all, are identified in the acknowledgements; a few prefer to remain anonymous given current or near-term resource planning activities. In total, the research team conducted interviews in-person or by phone with representatives from 13 entities in the electric sector, including nine utilities. The authors gathered information on utility supply planning and how solar is represented, including:

- Resource planning process details (planning horizon, expected net load growth),
- Key assumptions (whether DG is represented as supply or negative load, solar cost assumptions, existing solar projects)
- Model methodology (inclusion of distribution and transmission planning, type of risk analytics, candidate portfolio development),
- Model software (which models for capacity expansion and production simulation); and
- Solar representation (project size, capacity value, integration cost adder).

To supplement the interviews, the research team designed and administered a questionnaire. The questionnaire, which was distributed to 58 utilities using a web-based platform, included 25 questions focusing on the methods and tools used in the planning process and the incorporation of solar into this process. Key questions and a summary of utility responses from the utility questionnaire are included in Appendix B.[11] Twenty-two responses were received, representing 28 utilities in 22 states.[12]

1.6. Report Organization

In Section 2, the report provides background information, including the IRP/long-term planning process, laws and regulations that shape electricity supply and demand (e.g. state mandates for renewable energy and energy efficiency), and related research to date. Section 2 also covers details of the resource planning process and explains considerations about the existing generation fleet, load forecasts, capacity expansion planning, production cost modeling, and long-term portfolio selection. The inclusion of solar in resource planning processes is explored in Section 3. This section includes a solar PV technology overview, information on where to get solar data, experiences in capacity valuation, and additional considerations in how to include solar in resource planning. Section 4 explores challenges and benefits with including solar in the supply planning process that were identified by the utility participants and stakeholders, as well as some utility-identified answers for thirteen key questions around including solar in utility long-term planning. And finally, next steps and utility-identified analysis needs are discussed in Section 5.

2. RESOURCE PLANNING PROCESS

Integrated resource/long-term planning is a utility function that looks across a wide range of potential resource options to decide the best combination to reliably serve customers over a long-term period, often 20 years or more. This long-term view is often codified in an IRP or other long-term plan.[13] Based on interviews and questionnaire results, it appears that most – but not all – utilities follow a similar structure for long-term planning. It is also clear that for each step, utilities use different levels of detail, models or alternative calculation methodologies to analyze their generation portfolio. While resource planning is much broader than just the incorporation of solar, the differences were stark enough that the authors decided to capture the different methodologies used, as they provide a basis for incorporating solar into long-term planning.

2.1. Key IRP Metrics: Least-Cost and Lowest-Risk

Historically, IRPs were judged almost exclusively on a least-cost perspective, meaning utilities must choose whatever mix of future resources result in the lowest cost. Cost still appears to be the dominant metric guiding long-term planning decisions, particularly because the least-cost requirement may be codified in state law. With the advent of renewable energy and energy efficiency standards, coupled with several natural gas market disruptions and associated price spikes, some state commissions and utilities are increasingly looking toward incorporating supply diversity and risk-reduction into their planning model. For example, some utilities are placing an emphasis on risk-adjusted metrics that consider future regulations, future fuel prices, "affordable costs" or other non-cost metrics.

However, some states may require state law updates in order to consider least-risk in addition to least-cost. For example, the Wyoming Public Service Commission updated their guidelines for IRPs to specifically state that the commission's review of resource plans may include "least-cost/least-risk planning" (WY PSC 2013). Utilities who note that an exclusive focus on least-cost may not yield the optimal portfolio of resources, may decide to push for changes in state law and statute to allow for inclusion of least-risk considerations in addition to least- cost.

Incorporating regulatory and fuel price risk and variability into the planning process introduces significant complexity but provides a much fuller picture of how different assumptions regarding future resource mixes will impact key metrics for the utility in question. For example, the following questions can be more closely analyzed when considering *least-risk* in addition to *least-cost*:

- What is the impact of a 30% price swing in natural gas on the future revenue requirements of multiple portfolios?
- How do more stringent environmental regulations impact the economic viability of the existing coal fleet?
- Will higher load growth materially impact the resource decisions being made in the next five years?
- In 30 years, what are the potential risks and uncertainties with building resources today?

To determine the answers to these questions, the utility must examine relevant supply and demand policies, look at its existing fleet of generation resources, project the longevity and availability of each asset and contract, consider demand-side opportunities to reduce load and/or peak demand, forecast future energy and peak demand requirements, and determine the best set of future resource additions to meet customer needs. As a result, there are a large number of variables that could change over a 20–30 year planning horizon, and thus a large number of potential futures that could come to fruition.

It is important to note that utilities plan to meet two related yet distinct customer needs as part of this assessment: peak demand (in MW) and energy requirements (in MWh). From a peak demand standpoint, utilities typically plan to meet the forecast load growth plus a reserve margin. This reserve margin is often based on a Loss of Load Probability study, i.e., determining the reserves needed to ensure a probability of no worse than one outage in a 10-year period. Reserve margins vary slightly for each utility, but typically fall in the 12–18% range of peak demand (meaning that for every 1,000 MW of peak demand, utilities plan for 1,120–1,180 MW of generation availability).

From an energy perspective, the mix and type of each generation resource becomes much more critical. A utility could conceivably meet all of its needs by adding combustion turbines; however, these resources typically have a poor heat rate (or lower fuel efficiency), which translates into a high overall production cost to the consumer. Alternatively, the utility could meet all of its future needs with renewable energy, which effectively has no production cost due to its fuel source being free; however, renewable energy is often variable in nature and not perfectly aligned with system needs. This could result in a significant over-building of capacity than would otherwise be pursued. It is for these reasons that integrated resource planning is such a critical activity for so many utilities – the decisions contemplated in this process can have huge ramifications to customers over the long term.

2.2. The IRP Process

Integrated resource planning is the process of determining the lowest practical costs of reliably meeting expected future electricity load through the combination of supply-side generation and demand-side resources, while balancing reliability of service, environmental protection, and other goals.

Given the benefits of careful planning, particularly to ratepayers, some states require utilities to undertake IRP or long-term planning processes that are overseen by the state regulators. As of 2011, 27 states had IRP requirements, two states were developing or revising IRP rule or filing requirement, and ten states had long-term plan filing requirements (see Figure 1). The planning horizons generally range from 10 to 20 years, with the majority using a 20-year period. Plans must be updated every 2–5 years, with most states requiring biannual updates (Wilson and Biewald 2013). In general, state rules require that all feasible supply and demand side resources be considered, with some states specifying the resource types.

Long-term planning can be done differently, depending on the structure of the market and the authority of the regulators. In states with regulated retail markets (mainly in the West and South), generation suppliers are regulated and often have state requirements to produce an IRP or long-term plan. In states with competitive retail markets (mainly the Northeast, Mid-Atlantic, parts of the Midwest and California), the suppliers of generation are not regulated, and thus do not have state requirements that suppliers produce an IRP. Distribution companies that operate in competitive markets, however, are the "provider of last resort" and are thus responsible for procuring power for customers that do not choose a competitive generation supplier. These distribution companies are subject to oversight by the public utilities commission. As of 2011, four states with competitive retail markets (Ohio, Delaware, Connecticut, and Rhode Island) required distribution utilities to undertake an IRP process, although the resource plans are generally not as comprehensive as in states with vertically-integrated utilities (SEE Action 2011). Even in the states that do not require the submission of IRPs or long-term plans, utilities often undertake their own resource planning efforts as a matter of prudent business practice either on regular intervals or as needed based on market conditions.

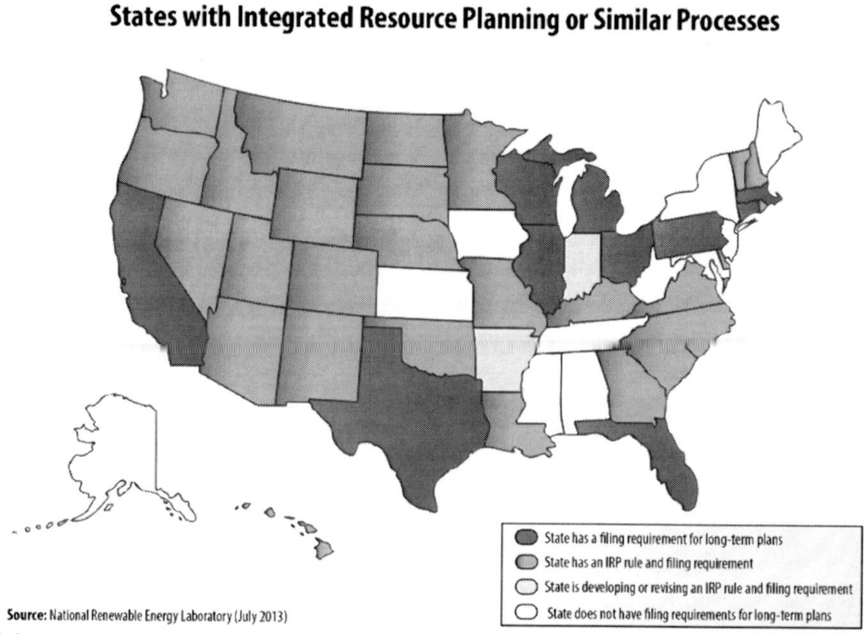

Adapted from Wilson and Biewald 2013.

Figure 1. States with IRP requirements as of 2011.

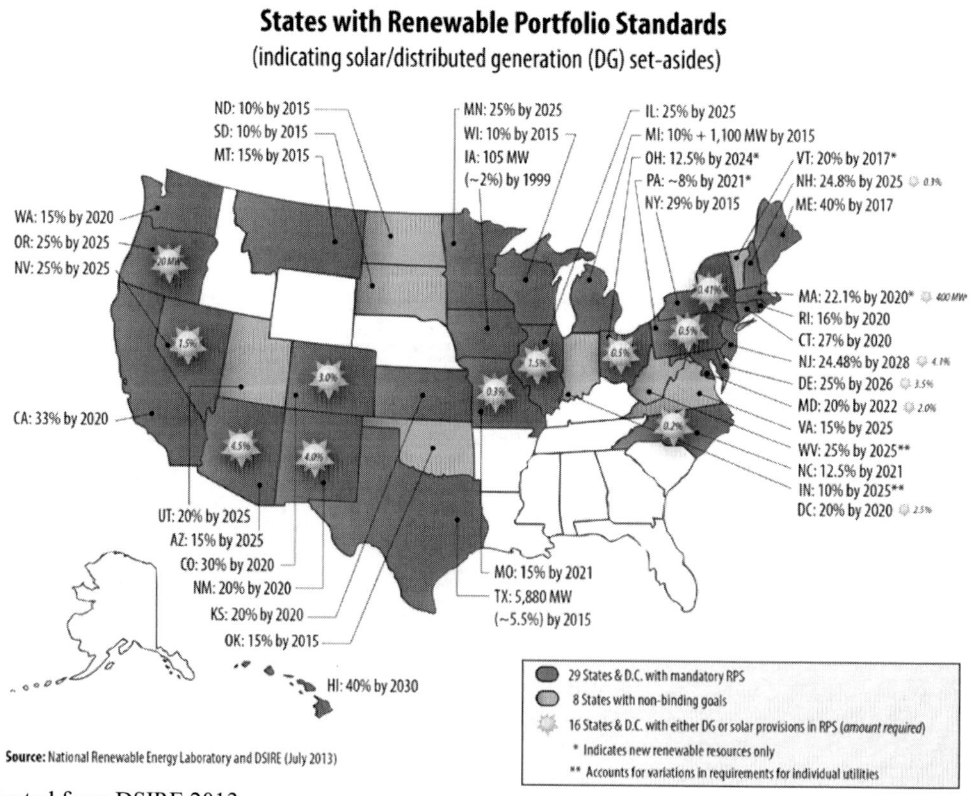

Adapted from DSIRE 2013.

Figure 2. States with RPS requirements.

2.3. Electricity Supply and Demand Laws and Regulations

Many states have laws or regulations that can impact the mix of generation supply, or the total amount of customer demand. The dominant supply policy is the renewable portfolio standards (RPS). RPS policies require electric utilities to generate or buy renewable energy (RE) equal to a certain percentage of their retail electricity sales, or a set minimum quantity, within a specified timeframe. Currently, 29 states and the District of Columbia have established RPS mandates, while eight states have less binding goals. RPS policies vary by state in their requirement, deadlines, types of utilities covered, eligible resources, penalties for non-compliance, and technology eligibility. Several states have specific solar or distributed generation "carve-outs," shown in Figure 2.

Together, current policies require 3–5 GW of new renewable energy capacity to be added each year between now and 2020, with a total of 94 GW of new renewables by 2035, if full compliance is attained (Barbose 2013). States sometimes adjust the standards, historically to increase the requirements, though several unsuccessful legislative initiatives to limit or cancel RPS requirements occurred in 2013.

Renewable energy mandates may be incorporated into resource planning by setting constraints in capacity expansion and production cost modeling and by including the mandate as a criterion during the evaluation of resource portfolios.

Demand-related policies can include energy efficiency mandates, utility energy efficiency programs, demand response/shifting and technologies such as smart meters. These programs are evaluated to estimate program growth and then rolled into the overall load forecast (see section 2.5) to determine what peak demand is and when it will occur.

2.4. Existing Generation Fleet Review

The first step in resource planning is determining the longevity of the existing fleet of generation assets and purchased power contracts. Generation can be categorized into four main categories: baseload, intermediate, and peaking. Baseload generation assets are typically larger plants that have very low operating costs and are intended to run on a nearly constant basis. They also require a long lead time to start and therefore ramping on and off are not particularly viable for these facilities, particularly as it could have an impact on minimum loading requirements for the generators. Nuclear and coal generation are often treated as baseload resources. The degree that baseload facilities can ramp up and down is dependent upon the design of each individual plant. NREL research shows that repeated cycling of coal plants can have large impacts on cost and wear (Lew et al. 2012).

Intermediate resources, typically natural gas combined-cycle plants, dispatch next. These resources are typically more flexible (compared to baseload plants) and able to follow load during the day. They can also be started more quickly than baseload resources (in a few hours), can be cycled on and off daily, and can be ramped up and down relatively easily.

Peaking resources such as combustion turbines are used to meet demand during the peak hours of the day when load is at its highest. These assets can start and stop very quickly (often in 10 minutes or less), but are more expensive to operate. Graphically the three dispatchable resource types can be depicted as serving distinct portions of a utility's load duration curve, shown in Figure 3.

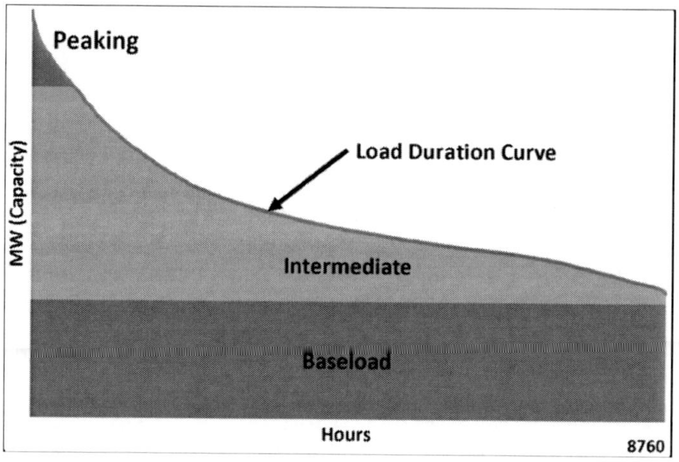

Figure 3. Load duration curve by resource category.

Certain renewable resources can be placed into a fourth category called "variable." Renewable resources like solar and wind are not dispatchable; that is, they cannot be turned on and off at the utility's discretion. Moreover, they are effectively "must take" resources, meaning that when the sun shines or the wind blows, the utility will be taking the energy

Separate from utility-owned generation assets, many utilities rely upon purchased power agreements (PPAs), power exchanges, storage, and short-term market transactions to meet a portion of their load requirements. Contracted power could be in the form of fixed blocks (100 MW delivered for a given time period), tolling agreements (where the utility is responsible for the delivery of gas and dispatching the third party's unit), or call option contracts (executable when the power price exceeds an agreed upon strike price). PPAs differ significantly from owned assets in that when their contract term ends the resources must be replaced or renegotiated, whereas with owned assets there is the potential to extend the plant's life.

When approaching long range planning, the utility must understand what tools it has in the proverbial toolbox and how long those tools will be accessible. As infrastructure ages, more and more generation assets will be nearing the end of their useful life and plant retirement decisions will loom large. This is particularly relevant with existing coal-fired generation, where environmental regulations requiring expensive upgrades to emissions control equipment, coupled with low natural gas prices, cause resource planning groups to perform detailed economic analyses of repair-versus-replace scenarios. Plant retirements could require the addition of new generation assets regardless of whether the utility forecasts any customer load growth during its planning horizon.

2.5. Load Forecast

A key component in long-term planning is predicting future customer demand. This involves detailed analytical reviews of each customer segment and the likelihood of each segment adding new customers as well as increased (or decreased) consumption on a per-customer basis. Residential customers can change their electricity needs without physically changing their footprint in several ways. Increased load can result from more computers, smart devices, and video games, as well as replacing old technology with new, higher energy-use technology (e.g., plasma TVs). Decreased load can occur when appliances are replaced with more energy efficient models. For each customer class, the utility rolls up the load forecasts to determine the peak demand periods on a daily, monthly, seasonal and annual basis.

According to data collected by the Energy Information Administration, electricity sales growth has slowed significantly over the last several years compared to the previous several decades (EIA 2013).

Utilities as recently as five years ago were building resource plans based upon compound growth assumptions of more than 2% per year. The economic slow-down that began in the late 2000s is a principle cause of the reduced demand growth in recent years. In addition, many utilities have implemented energy efficiency programs and goals, which further reduce energy and demand projections. Others have implemented programs designed to reduce demand (utility energy efficiency programs), or to moderate demand including the use of smart meters, demand side management and programs designed to shift or shed peak loads. Based on a questionnaire conducted in support of this report, utilities appear to have lowered their overall internal load growth forecasts and are now more in line with what EIA is predicting (see Figure 4).

Table 1. Historical and Projected Load Growth

Time Period	Average Annual Load Growth
1981 – 2005	2.3%
2006 – 2012	0.23%
2013 – 2040 (est.)	0.78%

Source: EIA 2013.

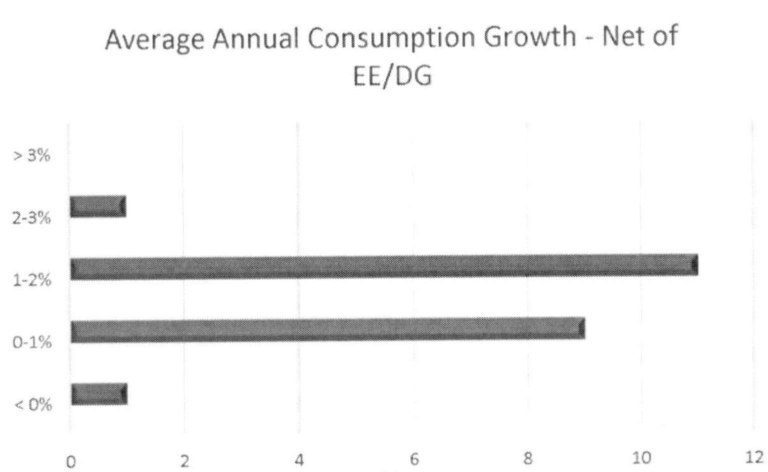

Source: Utility questionnaire, as part of this project[14].

Figure 4. Average annual consumption growth (Net of EE/DG).

Aligning a growth assumption with the existing generation and purchased power assets gives the resource planning group a vision of how much new generation capacity is required and when those additions are needed. This task is tackled as part of capacity expansion planning.

2.6. Capacity Expansion Planning

Capacity expansion planning is the act of determining the most viable long-term resource plan that meets projected electric demand. The process includes reviewing a wide array of new resource additions or contract renewals to meet growth or fill needs created by asset retirements and purchase contract expirations. Multiple future energy resource mixes, based on a wide array of potential options, are considered in aggregate. As mentioned previously, many utilities have operated or continue to operate in a least-cost planning environment; the determination of what is truly least-cost is a difficult task. Utilities have at their disposal dozens of technology choices and multiple configuration options within those choices. Figure 5 shows some of the choices for natural gas and solar.

For each viable technology option, the utility will track and calculate a series of cost and performance characteristics that will drive decisions about whether the resource will be included in any year of a future potential resource plan. Representative characteristics are shown in Figure 6.

Based upon these options and their unique characteristics, a plan for future resource additions can be developed. Two general forms of capacity expansion planning occur today: capacity expansion modeling and engineering judgment.

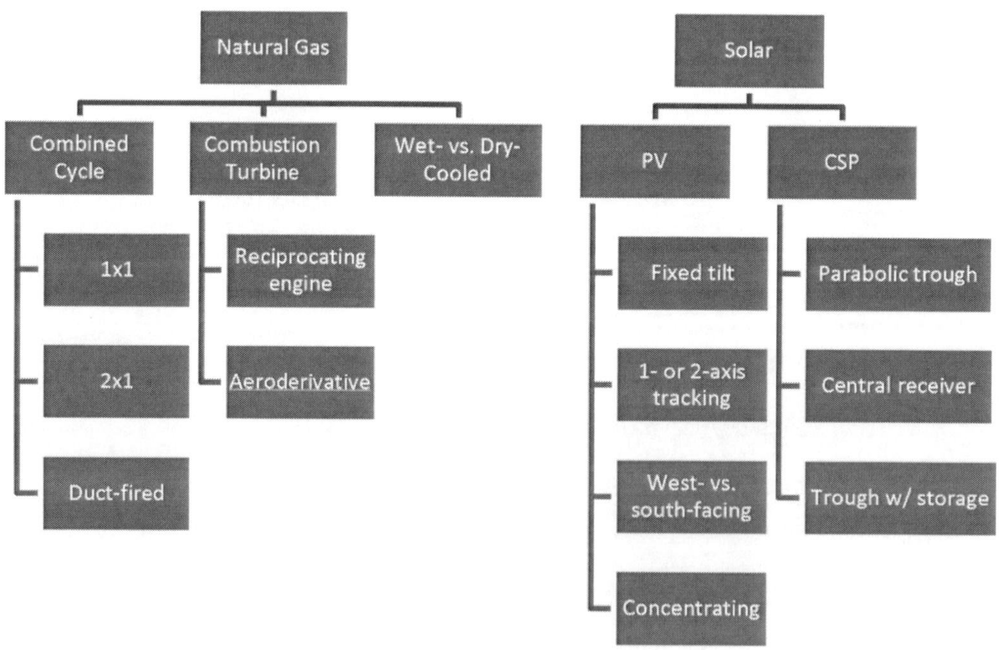

Figure 5. Sample resource selection options.

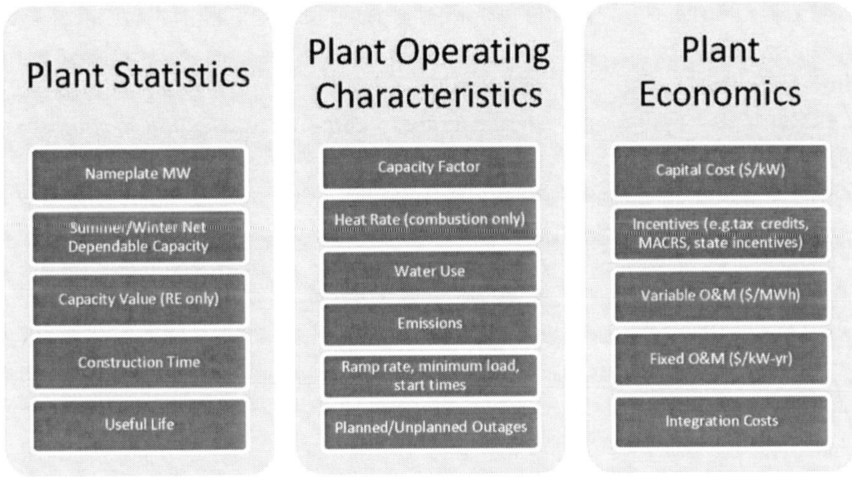

Figure 6. Resource characteristics used in capacity expansion planning.

Table 2. Examples of Capacity Expansion Model Constraints

Potential Capacity Expansion Constraint Criteria	Motivation
Limiting the number of specific resources that can be added in a given window of time	This approach is taken if the utility knows that it will be constrained from a capital or workforce perspective and could not build more than a given number of resources (i.e., combustion turbines) in any given year or over a set period of time.
Setting a minimum level of capacity or energy from a specific resource type	If a utility is subject to an RPS standard, this approach could be used to force a renewable resource to be selected even if it is not the least-cost option.
Restricting certain resources from being selected	A utility may wish to run a scenario where the model is not allowed to select certain resource types, like new coal generation, or limits how much capacity can be from PPAs.
Requiring a specific resource to be built at a certain point in time	The utility could force the model to place a specific unit in service during the planning horizon. This could be done if the utility wished to see the impacts of adding new nuclear capacity.
Forcing a plant retirement prior to its book life	Certain environmental regulations could warrant early closure of coal-fired generators. While capacity expansion models often can run "repair versus retire" analytics, the utility may wish to see the retirement scenario play out regardless of the least-cost option.

2.6.1. Capacity Expansion Modeling

Capacity expansion modeling is a process of creating a series of resource plans using a computer software tool called a capacity expansion model (CEM). Utilities provide an array of inputs into the model with details on their existing generation assets and purchased power contracts, load growth assumptions, cost curves for fuel sources, capital and O&M costs for potential resource additions, and other key variables. The models then typically run through a dispatch simulation (often a sample week in every month of the planning horizon) that reviews existing resources against load requirements and determines when new resources are required. The model then reviews the resource options available and identifies the best choice based on set criteria, often times using least-cost as the primary criterion (as noted earlier, some utilities are also including one or more lowest-risk criteria). To do this, the model looks at literally thousands of potential resource addition combinations and uses filters or "screens" to reduce them down to either one optimized plan or a ranked series of plans.

Certain constraints can be forced on the model so that multiple outcomes result, which can be further investigated with additional modeling. For example, Table 2 shows the potential capacity expansion constraint criteria, along with an explanation of how each criterion is applied.

By varying these constraints with each successive run of the CEM, a utility will be able to see the resource plans optimized to different end goals. Another way of thinking about this process is the concept of scenario planning. Several utilities use a series of "future states," or potential scenarios that could play out in the future, to drive how they could approach building out resource plans.

Scenario planning often tries to set plausible boundary cases based on existing unknown conditions. For example, one scenario could contemplate a future with extremely restrictive federal environmental policies. The utility could then look at constraining the CEM by forcing closure of coal plants, using high natural gas price curves, setting a carbon tax, or other activities. The CEM would then look to build a portfolio that operated at the least cost possible in that potential future, possibly by focusing on renewable additions. Alternatively, a scenario could be developed that postulated persistently low natural gas prices due to shale gas availability. This low-cost gas curve would likely lead to natural gas generation being the dominant new resource of choice.

The result of these constrained scenarios is a series of resource plans optimized to different assumptions and that outline the construction timeframes and availability dates of new resource additions throughout the planning horizon.

2.6.2. Engineering Judgment

One criticism of reliance on CEMs is that many of the resulting portfolios look quite similar. This is because the software is trying to optimize around cost given all of the constraints it is working against, and it will not make wildly different selections on its own.

Some utilities rely instead on engineering judgment to come up with distinct portfolio options. Developing portfolios using this approach often starts with an end goal and works backwards. A utility could look to create portfolios that met substantively different goals, such as a plan that would consume the least amount of water or a plan that used 30% renewable energy. These end goals could be internally generated or could result from gauging the interests of key stakeholders in the IRP process. Regardless, it is a viable method to determine the timing and addition of key resources.

2.7. Production Cost Modeling

The next stage in the long-term planning process is to run a production cost model (PCM). PCMs simulate hourly dispatch over the entire planning horizon rather than a representative sample as is done in a CEM. Utilities use these hourly dispatch models to perform complex sensitivity analyses. One approach is called Monte Carlo simulation, where multiple key input variables are randomly changed hundreds or even thousands of times in order to see how robust each potential resource plan is and how well it performs against a handful of key metrics. Other utilities will vary one variable at a time in order to see the specific impacts that major assumptions have on the outcome of a portfolio. Some utilities use both methods.

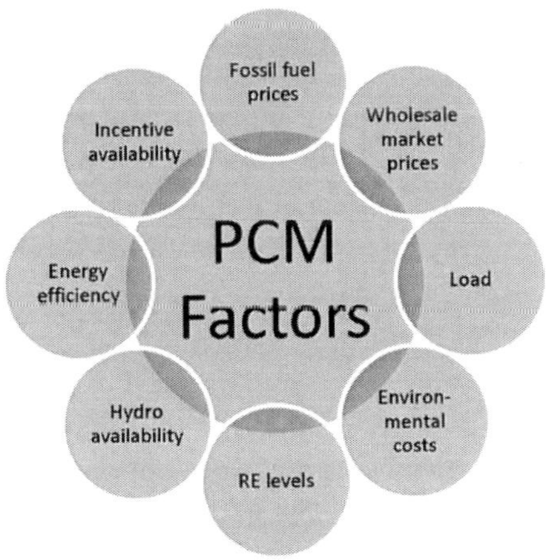

Figure 7. Common factors in PCM analytics.

Table 3. Common Resource Planning Tools and Their Associated Functions

Software Tool	Capacity Expansion Modeling	Production Cost Modeling	Transmission Simulation	Notes
Strategist	✓			Pairs with PROMOD IV
System Optimizer (Capacity Expansion)	✓			Pairs with Planning and Risk
PROMOD IV		✓	✓	Pairs with Strategist Regional scale Hourly load profile
Planning and Risk (PROSYM)		✓		Pairs with System Optimizer
AURORAxmp	✓	✓		
Strategic Planning (MIDAS)		✓		Used mostly for financial analysis.
EGEAS	✓			A modular package that optimizes to a load duration curve, with the single objective of least cost
GENTRADER		✓		Modular tool Regional scale Hourly load profile
PLEXOS		✓	✓	Models transactions to consider competition of supply, and is thus suitable for use in a competitive environment.

There are a group of assumptions utilities commonly vary as part of this process, as shown in Figure 7.

For multi-variable Monte Carlo simulations, variables can be correlated or allowed to move randomly. While capacity decisions may already have been decided, the changes in these inputs can align with the future state/scenario planning described previously. The utility could limit variables to certain ranges for each future state and test all of the portfolios against those cases to see how they would perform in a wide variety of situations.

By running an hourly PCM model, the utility is effectively looking at unit-by-unit dispatch across their system and instituting a dispatch of the resources based on economic merit. Modifications in certain assumptions can cause a significant shift in dispatch merit order. For example, adding a tax on carbon emissions reduces the economic viability of coal generation compared to other resources; depending on the level of the tax, it could result in coal resources being dispatched less frequently than would otherwise occur.

With this increased level of detail and precision, the resource plans developed in the capacity expansion planning phase can be compared with each other across a wide variety of sensitivities. This process can also give the utility an understanding of how resistant each resource plan may be to market unknowns and risks; this robustness (or lack thereof) factors into the decision-making process in many IRPs and long-term plans.

In order to simulate the complete picture, utilities often run these models in conjunction with each other, since capacity expansion and PCM tend to run as separate models. Table 3 provides a list of some common models currently utilized by utilities in resource planning.

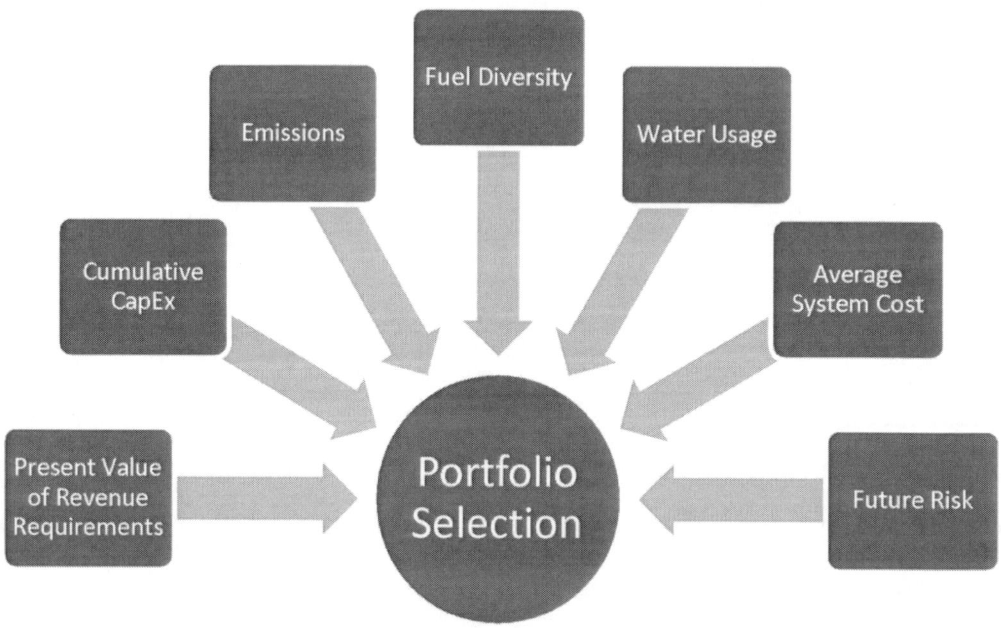

Figure 8. Key metrics for IRP and long-term planning selection.

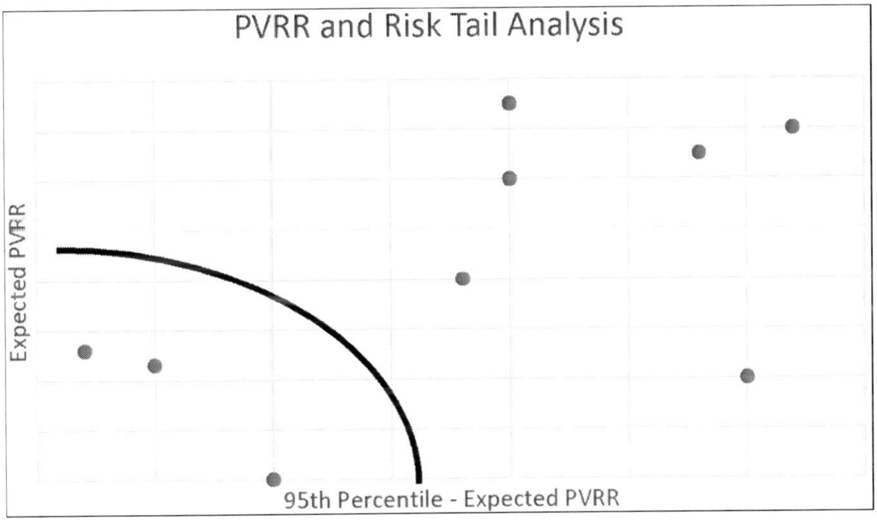

Figure 9. Sample scatter plot of portfolios for risk analytics.

2.8. Portfolio Selection

The end result for many IRPs is a recommendation on a resource portfolio. Selecting one resource plan from several (or dozens) created in the capacity expansion planning process and analyzed through production cost modeling can be a daunting task. To assist with this, utilities rely on a handful of key metrics to guide their selection. These metrics could be viewed qualitatively or quantitatively as part of a detailed scorecard. Metrics typically examined are shown in Figure 8.

Of particular note is the inclusion of risk parameters in the long-term planning process. Least- cost planning dominated the IRP landscape for many years; however, several states have introduced least-risk planning and metrics as a way to gauge the potential for long-term rate stability. These can include the value of generator diversity to lower potential future risks, as well as the potential risk of future regulations that could increase the risk of today's supply decisions. Risk metrics are often a result of detailed Monte Carlo simulations, where it is possible to create a bell curve of potential present value of revenue requirements (PVRR) for each portfolio. If a utility wanted to ensure that it selected a portfolio of future resources that was both low cost and resilient to upward price risk, it could focus on the expected PVRR as well as the potential price risk (or risk tail, represented by the upper 5% of PVRRs or some other value).

In the sample scatter plot in Figure 9, a handful of potential resource portfolios are plotted based on two key metrics: their expected PVRR, and the delta between their 95th percentile PVRR (i.e., upper tail of cost projections) and the expected PVRR. Portfolios falling in the lower left quadrant (as highlighted by the black line in Figure 9) would be considered to have lower expected cost as well as lower overall risk of price increases for ratepayers. While only a few utilities follow this approach, many have considered adopting this or some other similar method as part of their portfolio selection process. This can be augmented by also looking at the environmental impacts of each portfolio (carbon emissions, water usage, etc.), or even impacts to economic development criteria like job creation.

It is important to note that portfolios selected as part of an IRP process are rarely binding; rather, they are treated as roadmaps based on a given set of assumptions. Virtually all utilities have a separate process to receive approval for building new assets and bringing them into their rate base. Indeed, IRPs are often only "acknowledged" by state commissions, a term of art that provides little assurance to utility financial officers and investors, as it does not contain any level of approval or surety of future cost recovery; however, IRPs and long-term plans do provide an opportunity for the utility to lay out its vision of potential futures and the resource mix that is most appropriate for its customer base.'

3. Inclusion of Solar in Resource Planning Processes

3.1. Background

Solar generation has become increasingly prevalent in recent years, driven by an increased focus on renewable energy coupled with rapid decreases in the cost to deploy solar assets. Over the last three years alone, total annual solar installations have risen from 781 MW in 2010 to nearly 2,400 MW in 2012, and cumulatively nearly 6.1 GW of solar is now installed and operating in the United States as of the end of 2012 (Krishanmoorthy et al. 2013).[15] Three interesting characteristics exist for that installed capacity:

1. More than 99% of the installations are customer-sited (DG).
2. Nearly 80% of the cumulative U.S. solar capacity serves the load requirements of 10 utilities, with 30% of the overall capacity at one utility (Krishanmoorthy et al. 2013).
3. In the last 2 years, centralized solar (projects > 5 MW in size) have increased from less than 1 GW to more than 3 GW (SEPA 2013).

The proliferation of DG systems across multiple jurisdictions has been driven by many disparate factors. In some regions, it is due to aggressive customer adoption incentives. In others, the systems can compete on a cost basis against existing utility retail rates. Utility-scale systems, on the other hand, must compete more directly against wholesale costs, either all-source or renewable-only. The declining costs of PV in recent years have helped close the gap between what it costs to develop and deploy solar and what the market can bear. This became crystallized in 2012 when large-scale solar systems made up approximately half of all new solar capacity installed (slightly more than 1,100 MW), an increase of 2.5 times the capacity from 2011(SEPA 2013).

Not all utilities are adopting solar generation at the same pace; however, as solar deployment costs continue to decline and reach parity with other resource options (especially solar PV recently), more and more utilities are likely to turn their attention toward solar generation and how it can play a part in future resource plans. In fact, most if not all utility interviewees stated that they are likely to include solar PV as a resource in their forthcoming plans.

3.2. Solar Data

There are two main types of solar data required in long-term planning according to utility interviewees, which include solar profiles and solar costs.

3.2.1. Solar Profiles

Solar profiles characterize the potential solar generator output, which includes resource potential on an hourly basis. Utilities that participated in the research for this report identified three main sources for solar profile data that they incorporate into their IRPs/long-term plans: National Renewable Energy Laboratory's PVWatts (PVWatts 2013), PVsyst (PVsyst 2013), and data from existing operating plants in their service territory. PV Watts and PVsyst are only relevant for PV systems, the dominant technology included in long-term plans (if any).

PVWatts is an online program developed by NREL that provides the capability to create a typical meteorological year (TMY) of profile data for a solar resource in a given geographic region. This tool can provide a first glimpse into what kind of production could be expected from a proposed solar facility. Using a DC rating in kW, a derate factor, the type of array (fixed or tracking), and the array's tilt and azimuth (where south-facing is stated as 180°), the model will create an hourly expected output based on typical weather patterns and solar irradiance, as experienced over several decades' worth of recorded data. The downside to PVWatts is that the analysis can be run only for a limited number of sites. For example, there are only four sites available in Arizona: Flagstaff, Phoenix, Prescott, and Tucson.[16] This data can be particularly helpful if utilities do not have many PV systems in operation in their jurisdiction, or do not have access to those systems' output and want an extremely simple methodology.

PVsyst is a software package that allows users to study the design of PV systems and perform sizing and data analytics. It includes component and meteorological databases, and is intended for use by architects, engineers, and researchers. PVsyst performs three general levels of study: preliminary design, including sizing the project; project design, where detailed hourly modeling is conducted; and measured data analysis, which allows for the actual data from the fully operational plant to be compared against simulated variables (PVsyst 2013).

Data from operational plants is also used by utilities with access to system data within their service territory. This data provides the utilities a view into the real operational characteristics of solar PV and other solar generation. However, because these profiles are based on actual plant output, they may require some review and potential adjustments prior to inclusion in any kind of analytics. For example, if a PV plant has only been in operation for a short period of time, it may be necessary to supplement the profile with PVWatts data. Alternatively, severe or abnormal weather occurrences may dictate the need to augment with PVWatts data. This scrubbing is appropriate, as long-term plans are based first and foremost on projected load requirements which themselves have been "weather normalized," or stripped of any impacts from abnormal weather conditions. As solar output is also predicated on weather patterns, it is appropriate to ensure an "apples to apples" comparison.

While there is interest in the combination of storage and PV to improve the profile of PV overall, storage is not typically modeled in long-term planning by the majority of utility interviewees, because utilities feel they need more credible data and analysis before storage can be included in long-term planning analysis. A few utilities that do consider the combination of storage and PV consider the combination to have dispatchability, although on a limited basis.

3.2.2. Solar Cost Information

For the most part, solar cost information is collected from publicly available sources, and from existing plants in the utility territory. However, for PV technologies the market is changing so fast, it is hard for utilities to understand the future cost trajectory.

Utility interview participants universally commented on the rapid decline in PV costs over the last several years. In particular, most utilities had forecast a declining cost curve in their planning assumptions, only to see the actual costs decline much more steeply than anticipated. Common beliefs regarding the root cause of the nearly exponential cost reductions included:

- U.S. tax subsidies
- Foreign manufacturing credits
- Efficiency gains
- Market surplus of panels.

On a going-forward basis, there was still consensus but less overall agreement on where PV costs would trend. Companies that were questioned on this issue fell into three general groups of thinking shown in Figure 10.

Several utilities specifically mentioned the Department of Energy's (DOE) SunShot Initiative goals for reduced costs of PV deployment by 2020 as being potentially obtainable.[17] Others commented that (while the dramatic declines seen over the last 5 years are not sustainable) reductions will continue, albeit in a more measured fashion. Finally, several companies believed that solar costs were at or near a bottom and would levelize for the foreseeable future. These solar PV cost assumptions are reflected in their solar cost assumptions used in their long-term planning. While prices have come down rapidly, questions remain about how long the prices will continue to decline and when they will bottom out (or if they already have).

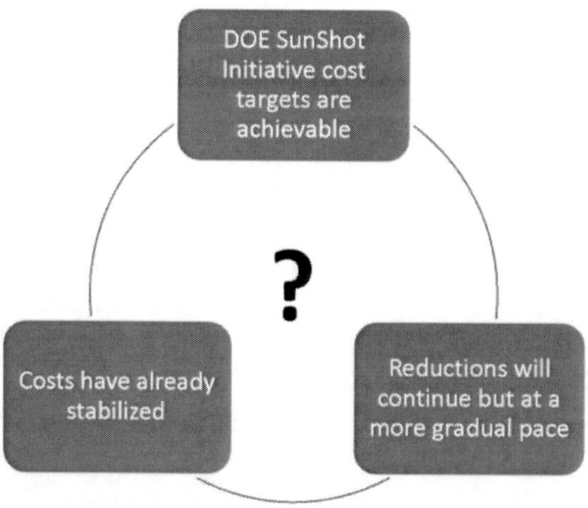

Figure 10. Opinions on future of solar costs.

A few items have contributed to the thinking of utilities that believe PV costs may have bottomed out. First and foremost, there has been a series of bankruptcies in the solar industry over recent years, which has taken several manufacturers out of the market. Second, demand for solar panels globally is increasing and the market surplus is expected to dissipate over utility planning horizons. Finally, there is concern over post-2016 when the ITC reverts from 30% back to 10%; demand could potentially slow.

Publicly available data sources for solar cost data that can be used in utility analysis are explained in Section 4.2.1.

3.3. Solar Capacity Valuation

One key question discussed at length with interviewees was the capacity value utilities attribute to the addition of solar capacity on their system. In essence, a solar generator's capacity value is the percentage of its nameplate capacity that is anticipated to be reliably available to meet daily and seasonal peak demand, and is a very region-specific metric. There are multiple terms of art for this occurrence, capacity credit and capacity value being two more commonly used.

Conventional generation resources, like natural gas combustion turbines and coal-fired plants, are fully dispatchable and can be called upon to meet load as it fluctuates throughout the day and year. In contrast, the most commonly deployed solar technology, fixed-axis solar PV, generates only when the sun is shining.[18] Typically, the sun provides the most energy for production between noon and 1 p.m. each day, and the production capability can diminish rapidly as the sun sets for projects facing south. PV facing west can moderate this drop-off, although overall production of the facility will be lower. Most utilities commonly peak later in the afternoon, between 4 p.m.–6 p.m., due to air conditioning load in the summer months. For this reason, solar energy may not correlate perfectly with a utility's need for it, depending on their specific situation (e.g., latitude, insolation). Different solar technology configurations can moderate this potential for rapid impact through the addition of storage, or PV tracking (one or multiple axes) – but these configurations are not always included in long-term planning modeling.

Several methods are used to calculate the capacity value. Some of the more commonly used approaches are explained in Table 4.

The appropriate capacity value for a given solar resource is dependent upon its generation profile and the utility's load that it is intended to serve. Fixed-tilt systems that are facing south, while producing more energy, will have a lower capacity value than similar systems facing west, toward the setting sun. Similarly, single-axis tracking systems will have a higher capacity value (and capacity factor) than fixed tilt because of their ability to follow the sun. Figure 11 shows results from a recent LBNL report that outlines the wide array of solar capacity values currently in use, as reported in utility IRPs (Mills and Wiser 2012). As shown, dispatchable technologies were granted the highest capacity credit/value (well over 80%), CSP (no storage) and PV with tracking were granted capacity credit/value in the range of 50–85%, and fixed PV was given a credit/value of 0–50%. General PV had the widest range, between 5% and 100%.

To build upon LBNL's work, the authors conducted direct interviews and gathered questionnaire responses. Most utilities interviewed in this effort generally utilized capacity values in the 20–60% range, depending upon technology selection. Adding storage capabilities or some other form of dispatchability generally would increase the assumption to 60–100% capacity value. However, several utility interviewees stated that they do not provide any capacity value to solar, with or without storage capabilities.

As part of the questionnaire issued in support of this report, utilities provided additional data about their internal planning assumptions, as shown in Figure 12. For different configurations of solar PV (first three technologies), 60–76% of utility questionnaire respondents assigned a capacity value between 0–60% for fixed PV and up to 80% with tracking. 24–40% of responses indicated no value assumed for PV technologies.[20] Of utilities that did assign capacity value, the greatest value was assigned to utility scale tracking axis PV; more than half assigned a capacity value greater than 40%. This makes sense because PV with tracking capability can help meet early evening peak demand and provide more capacity value overall. Fixed axis PV was allocated a lower capacity value than tracking overall – with a slightly higher value for utility-scale compared to distributed PV.

Table 4. Primary Capacity Valuation Approaches[19]

Method	Definition
Effective Load-Carrying Capability (ELCC)	Represents the ability of a power plant to increase the total load of a local grid without increasing its Loss of Load Probability (LOLP).
Peak Time/Season Windows	Calculates capacity credits across predefined hours, months, and/or seasons. This approach can look at either the minimum or median output likely to occur with a given probability. It does not account for the impact of different levels of grid penetration.
Capacity Factor	Quantifies a power plant's average output relative to its installed capacity; while this approach is simple to calculate, it bears no relation to load served, and cannot account for the impact of different levels of grid penetration for solar as a whole.

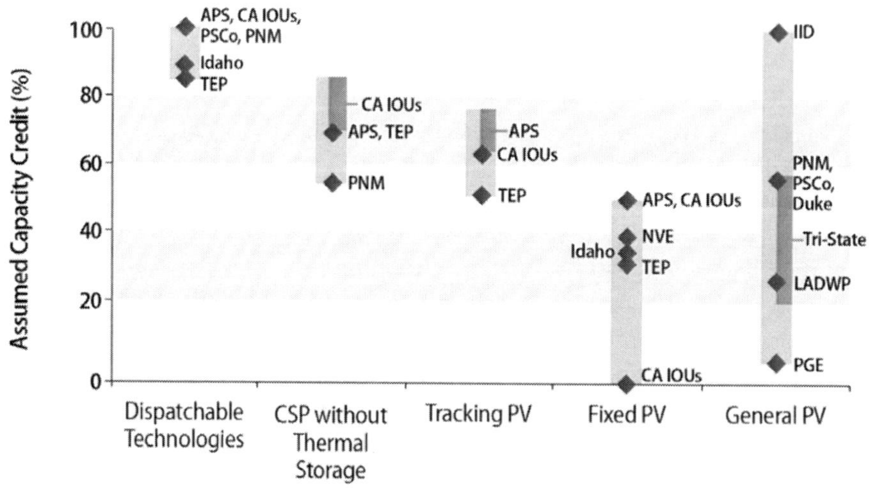

Dispatchable technologies include CSP with thermal storage, solar thermal gas hybrid, and PV with battery.

Source: Adapted from LBNL 2012.

Figure 11. Range of capacity values used in IRP plans, by solar technology.

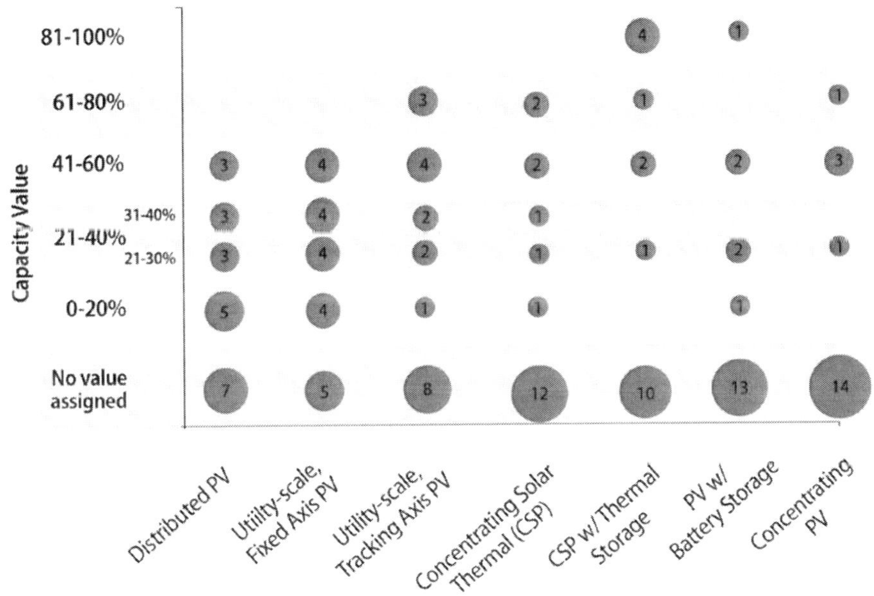

Note: Numbers in circles represent the number of utility responses.

Figure 12. Range of solar capacity values used in planning analysis, by technology, provided by utility questionnaire respondents.

Conversely, only 26–44% of utility respondents assigned any capacity value to CSP (with or without storage), PV with battery storage or concentrating photovoltaics (CPV); 56–74% used a 0% capacity value.[21] Utilities that assigned capacity values used a range of 0–100%. Half the respondents that assigned value to CSP with storage assumed a value of 80% or above and 88% assumed a capacity value of 40% or greater. For CPV technology, of those that assigned value, 80% assumed a value of 40% or greater.

When looked at in aggregate, the incremental capacity value for PV generation resources can actually decrease as penetration increases. This occurs because a utility must still meet system summer demand after the sun goes down around 7 p.m.–8 p.m. As its penetration increases, solar PV's ability to provide any relief to system needs diminishes rapidly after 5 p.m., as shown in Figure 13. It should be noted that this decrease does not become marked until a non-trivial level of penetration is reached (Perez et al. 2006). Five utility respondents to the questionnaire indicated they factor in declining capacity values for solar PV in their planning process; however, others were aware of the issue and indicated that it would be studied in the future if penetration levels warranted such an analysis.

Figure 13 depicts a typical utility load curve, with increasing levels of PV penetration netted out of the load. The solar peaks in the middle of the day, resulting in the full nameplate level of PV impacting the load. As the day goes on, the solar drops in production providing less and less generation coincident with the utility's typical peak hour.

Over half of the utilities participating in this report indicated that they perform their own analytics comparing solar resources in their region directly to their own load profile or were basing their capacity values on data from systems currently in operation. The remaining companies rely upon a generic value (either from an industry report or a consulting firm), or use the same capacity value for solar as they do for wind generation. Several utilities indicated that they were looking at developing their own capacity valuation for solar generation in the near future. Many utilities noted a desire to have access to a larger data set including other utilities – to compare their experience or have a good place to start in lieu of their own data. Appropriate and accurate capacity values are important in the long-term planning context because they will allow for solar generation to meet at least a portion of future demand requirements, thereby deferring the need for some new fossil fuel generation.

3.4. Additional Considerations for Solar in Resource Planning

3.4.1. Solar Integration Costs

One risk associated with solar generation without storage is adequately integrating its production fluctuations into the overall resource mix for a given utility. Based on responses to the questionnaire, discussions during interviews, and industry research, one commonly used approach is to estimate the cost (in $/MWh) to the system at large that was caused by the variable renewable resources. These studies look at the operating reserve requirements and what is needed on a 5- or 10-minute system operation timeframe. While the cost estimates are typically low, often in the $2–11/MWh range – as indicated by utility IRPs investigated by LBNL (Mills and Wiser 2012), they do represent a cost to the system once these resources are operating.[22] Utilities can incorporate this cost into their CEM to provide a more robust view of the costs of this variable resource.

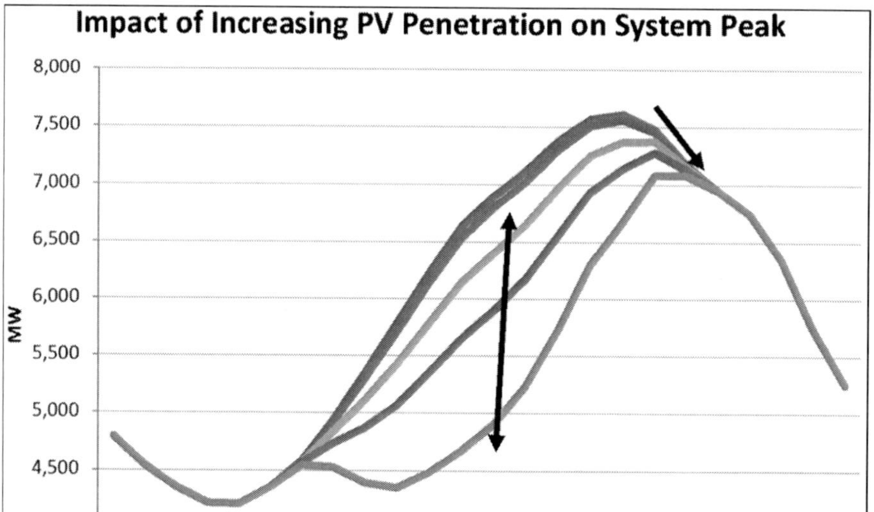

Note: HE denotes hour ending.

Figure 13. Impact of increasing PV penetration on ability to meet system peak.

3.4.2. Customer-Sited Distributed Generation

DG systems are typically installed behind the utility's meter. Because the system operator does not have complete control over DG, and because the production is generally used to reduce total site demand for utility power, most utilities today treat DG as a net load impact rather than as a resource (see Appendix B, question 3). This net load approach can happen implicitly or explicitly, by either assuming that the lower load growth projections that are now used already incorporate some assumption of DG behind the meter or by projecting a DG adoption curve and then subtracting it from a gross load growth projection for their overall customer base.

Alternatively, a small number of utilities treat DG (or are strongly considering treating DG) as a generation resource option, rather than a net load impact. This approach can allow the CEM to determine a more optimized approach to the amount of DG that could be included in the resource plan, and allows for independent investigation of customer load profiles (a major driver of long-term planning results). Some potential solutions for addressing DG in modeling can be found in Section 4.2.7.

3.4.3. From Solar Planning to Procurement

Resource planning focuses on long-term visions for where the utility wants to take its resource mix. The capacity expansion plan that results from the capacity expansion analytics performed in support of an IRP outlines the timing and type of resources that will likely be pursued. It then falls upon the power procurement division of the utility to turn those plans into projects that deliver energy and capacity to the grid. This section focuses mostly on PV procurement.

While long-term resource planning is used to inform power procurement at many utilities, these two operations are often run by two different departments within the utility. And the level of coordination between these groups can vary significantly from utility to utility.

Utility procurement procedures typically start with a request for proposal (RFP) that is generally targeted to a specific need, be it meeting an RPS goal, adding additional peaking generation, or replacing an expiring PPA. Some utilities may use the same/similar key metrics from the IRP reviews to inform the RFP process to determine the best mix of future resources. As discussed above, PVRR is the main metric on which utilities focus. Once an RFP is issued, the focus shifts from comparing large portfolios against one another to comparing specific projects on a case-by-case basis. One approach is to compare the proposals to an avoided cost proxy. At this stage, it is critical that the utility be able to model the specifics of each proposal to determine their relative ranking. And some utilities who want to have their IRP processes drive their procurement will review the RFP criteria and outputs with the IRP planning team to make sure the procurement process is lining up with the planning process results.

The main variable in solar bids will be the site-specific production profile. Where many utilities may rely today upon generic TMY data for modeling solar in their long-term planning analysis, each proposal submitted as part of an RFP will, by necessity, contain much more granular information that impacts the solar profile. Site footprints, technology selections, array orientations, and site irradiance levels all aggregate together to produce a specific solar profile that must be analyzed. As part of the RFP process, the utility can request expected production profiles that can be evaluated and compared against each other. In fact, many utilities keep a database that compiles a historical record of all of the responses to solar RFPs; this is used as a baseline to compare against future bids.

Some renewable developers have begun to study utility valuation methods and preferences to create tailored proposals. Often, these unique proposals may have a higher development cost or lower production capabilities, but actually provide greater value to the utility reviewing the bid. As mentioned previously, a system facing west will not produce as much energy as one facing due south, but the production profile could line up much more favorably with the utility's system peak and therefore offer greater capacity value (boosting its overall value to the utility).

From an analytical perspective, reviewing individual bids as part of an RFP process is mechanically similar to the analytics performed in the IRP process. The utility is still comparing the resources and looking for the overall lowest cost and/or best value for its customers.

By integrating more solar into the IRP process in general, utility staff will be more familiar with the modeling characteristics associated with solar generation and be more adept once bids begin coming in for those resources. Over the next several years, assuming solar development costs continue to decline and/or existing supply costs increase, more utilities will see solar generation approach grid parity with conventional generation options. It is anticipated that this trend will first become noticeable as part of traditional RFP processes, at which point those reduced pricing trends will filter back into routine long-term planning processes. Reductions in solar costs could also drive CEMs to begin selecting solar resources above and beyond what may be required to meet RPS requirements.

4. Improving Solar Within Utility Supply Planning

A number of key challenges of incorporating solar into supply planning analysis were raised during the utility interviews. Most of these utilities have not experienced notable levels of solar PV, CSP or CPV penetration, and have not always modeled solar in their IRP studies. Most utilities expressed an interest in incorporating and refining distributed- and utility-scale solar PV in their modeling[23] because they anticipate more PV coming onto their system in the next decade. Some utilities noted that CSP and CPV are not appropriate technologies for their service territories. Several utilities mentioned that they participated in this project specifically to learn from their peers about best practices for including DG and utility-scale solar PV in their models.

This section explores some of the key benefits and challenges of incorporating solar PV into supply planning analysis and decision-making raised during the utility interviews. Note that this list is not comprehensive. It captures only items/characteristics raised by the interviewees; others may exist or emerge. This section also provides questions and answers to these challenges – many of which were identified directly by the interviewees. The goal of capturing this dialogue is to help advance the discussion of how solar could be better represented in both utility supply planning and broader utility operations.

4.1. Utility-Identified Benefits and Challenges

During the course of interviews with utilities and stakeholders, several benefits and challenges associated with solar were brought up and discussed. These are not meant to be all-inclusive or applicable to all utilities. Rather, they give a sense for how solar generation is currently perceived, which may lead to a stronger understanding of how it could be adopted in the future.

4.1.1. Solar Benefits

- ***Meeting renewable standards requirements*** – Many states have renewable standards in place, and several of those have specific requirements for solar energy and/or solar renewable energy certificates (SRECs). Several utilities identified renewable mandate compliance as a key factor in their pursuit of solar generation.
- ***Fuel diversification*** – Many utilities are beginning to look at fuel diversification as a way to mitigate rate volatility for their customers. Utilities that are coal heavy expressed concern about future environmental regulations, while those that rely upon natural gas generation viewed severe historical price volatility as a motivating factor to diversify their fuel mix.
- ***Cost stability*** – Solar energy has effectively a zero dollar dispatch cost, as virtually all of the costs to operate the facility are fixed (i.e., construction cost and some fixed O&M). Utilities were drawn to avoid fuel price-related risks by essentially guaranteeing a portion of its energy requirements will be met with a very stably priced resource.
- ***Geographic dispersal benefits and PV modularity*** – Solar PV's variability can be mitigated by dispersing solar projects across a wide geographic region. Utility interviewees liked that solar PV can be deployed on a very incremental basis; there is no standard footprint or capacity size that must be considered. PV generation is immensely flexible in this regard as it can be sized on a scale of hundreds of kilowatts to hundreds of megawatts. By deploying solar in smaller amounts across a wider region, a utility can smooth out any site-specific cloud variability and related quick ramping up and down. Targeting specific geographic locations for PV installations can also allow the utility to solve localized voltage concerns or avoid transmission curtailment issues within load pockets, where siting generation assets (particularly those with emissions) can be problematic. PV generation also benefits in its substantial difference in construction schedules compared to conventional natural gas and coal. Solar PV facilities can be constructed and brought online in as few as six months, while natural gas resources can take anywhere from three to five years.
- ***Partial correlation with peak demand*** – Utilities' summer peak demands are driven primarily by air conditioning load, which results in system peaks in mid- to late afternoon each day, when the sun is still in the sky and solar generators are still producing some energy. The correlation of solar production to peak demand varies with each utility, location, technology selection, and orientation (Perez et al. 2006). Utilities interviewed for this report repeatedly stated that they believed there was a capacity value associated with solar – while not always 100%, it can be higher

capacity value than with wind generation. This subject was covered in more detail in Section 3.[24]

- ***Mitigation of environmental compliance risks*** – Solar energy is produced with zero emissions of CO_2 or other greenhouse gases (GHGs). With the federal government considering a carbon tax, cap-and-trade system, or some other mechanism intended to curb GHG emissions, utilities are contemplating future compliance costs. Diversifying portfolios with solar generation provides the added benefit of displacing emissions-generating resources.
- ***Avoided line losses (typically DG only)*** – DG resources are located at the load (the end user of electric power) rather than on the high-voltage transmission system like other forms of centralized generation. Because it is generated near the end user, the energy does not have to be transmitted across transmission and distribution lines, nor does it need to undergo any transformation (step-up or step-down) to serve load, meaning all of the typical line losses associated with remote, centralized generation resources are saved. Utilities liked that small-scale central station PV can also be located near load centers and interconnected at distribution voltages, avoiding line losses in a similar manner.[25]

4.1.2. Solar Challenges

- ***Variable and Uncertain Output*** – Solar is a variable resource that ramps down (e.g. cloud cover or storms) and is non-dispatchable. This means that utilities treat it as a "must take" resource and modify their economic dispatch decisions around its profile or production curve. The variability of solar is difficult to predict in advance and can lead to sudden drops and rises in the production profile at a given location. Utilities account for this variability by maintaining both spinning reserves and quick-start natural gas capacity. One third of the utilities responding to the questionnaire indicated that they include a cost add-on when reviewing variable resource proposals to account for the additional ramping requirements placed on their natural gas fleet.
- ***Ramping issues*** – With sufficiently high penetration levels of solar generation, operational issues can arise due to the solar generation ramping up rapidly in the morning, as the sun rises, and then ramping down rapidly in the afternoon as the sun sets. These severe ramps require the utility to have a large amount of flexible resources available (CA ISO 2013).[26]
- ***Economics*** – Utilities operating in a least-cost planning state must be able to prove that solar generation is economic for its customers compared to other available resources, either all-source or renewable-only. In regions where the solar irradiance levels are high, like the desert Southwest, large-scale solar has begun to compete on an economic basis with conventional natural gas-fired resources; however, in other states this convergence has not occurred. Utilities in least-cost planning paradigms that choose a more expensive resource are at risk for cost disallowances, unless they are assured there will not be a prudency review, or some form of cost recovery is pre-established. Utilities also mentioned difficulty in pursuing solar now while prices are still in steep decline, and the consequences of locking in prices that are significantly higher than what could be available 1–2 years later.

- ***Lack of current capacity need*** – The economic downturn experienced over the last several years has not only stalled load growth for utilities; in many cases it has actually caused a load reduction. Because utilities plan several years in advance to meet load, this change from the predicted load growth left many utilities long on generation capacity. In states with RPS requirements, additional generation capacity is being pursued regardless of overall capacity need, mainly because utilities want to avoid paying RPS non-compliance penalties. Adding solar generation in states that are not subject to an RPS could result in prudency concerns and associated cost disallowances.
- ***Cross-subsidization concerns (DG only)*** – When solar is installed behind a customer's meter, the generation from that system offsets the customer's electric usage. When that usage is less than the solar production, the generation gets exported to the utility's distribution system. This transaction is often completed under a rate paradigm known as net energy metering (NEM). The concern from utilities comes from the fact that most retail rates (particularly those for residential customers) are designed to recover both fixed and variable costs primarily through consumption-based variable rates. As DG is installed behind the meter and offsets (or reduces) energy and load requirements, these fixed costs can go under-recovered. This issue is largely separated from utility planning requirements, but was mentioned during interviews as part of the research for this report.
- ***Reduced capacity benefit over time with increasing solar penetration*** – As solar penetration increases and begins to have a larger impact on the system in aggregate, the overall benefits that each solar system provides decreases. Utilities are concerned about this decreasing capacity benefit. This concern is expected to rise unless utilities are rigorous about adequate dispersal of resources or energy storage solutions become commercially viable from a cost and technology perspective. Potential analysis solutions are explained in Section 4.2.2.

4.2. Utility-Identified Key Questions and Answers

Utilities and stakeholders identified their key questions about including solar as part of their long-term planning and analysis activities. Some even mentioned potential innovations or analysis methodologies to get answers to these questions. Both the key questions and answers that emerged as themes in our conversations are captured here, as best practices. Note that this section is not meant to be all-inclusive or applicable to all utilities.

4.2.1. What Are Credible Estimates of Future Solar PV Prices and Performance?

Many utilities acknowledged that in the last few years, solar prices are declining in a non-linear fashion, and a few mentioned that it appears the market will reach the U.S. Department of Energy SunShot cost goals. However, collectively, utilities had many questions about how long the solar prices will continue to decline rapidly and when prices will bottom out.

The challenge of predicting future solar PV prices was seen as a key uncertainty by most interviewees, particularly because planning is done over a multi-decade horizon. In order to consider the full range of potential resources that will be available, an accurate depiction of the trajectory of future solar PV prices would be extremely helpful.

Research and utility interviews revealed a number of reasons why PV prices are hard to estimate. There are a number of different references and data sources, and they are not always in agreement about the future of solar PV prices. It does seem that the further out in time you go, the fuzzier the price of PV becomes and the wider the range of predictions grows. There appear to be geographic distinctions in terms of resource quality, and experience of installers and other cost factors that aren't clearly articulated in the literature. And there were questions about the current market oversupply of PV – when will it be exhausted and the supply-demand balance be reestablished? Could PV prices increase once the oversupply is exhausted, and if so, by how much? Together, these factors make PV prices seem hard to predict.

NREL and SEPA both strive to create credible, objective analyses on a variety of topics, including the cost and performance of solar PV. As such, both entities have assembled a list of useful datasets, information, tools, and reports that could help inform solar PV predictions. They include:

Data Sources that are Regularly Updated

1. NREL's Transparent Cost Database –utility scale data (http://en.openei.org/wiki/Transparent_Cost_Database)
2. NREL's Energy Technology Cost and Performance Data for Distributed Generation (http://www.nrel.gov/analysis/tech_cost_data.html)
3. NREL's Open PV Project, which captures voluntary historical cost and performance data (https://openpv.nrel.gov/)
4. LBNL's *Tracking the Sun* report (Barbose et al. 2013 – http://emp.lbl.gov/sites/all/files/LBNL-5919e-REPORT.pdf)

Data and Analysis Snapshots in Time

1. U.S. Department of Energy November Technical Report: *PV Pricing Trends: Historical, Recent, and Near-Term Projections* (Feldman et al. 2012 - http://www.nrel.gov/docs/fy13osti/56776.pdf)
2. U.S. Department of Energy November Technical Report: *Benchmarking Non-Hardware Balance of System (Soft) Costs for U.S. Photovoltaic Systems Using a Data-Driven Analysis from PV Installer Survey Results* (Ardani et al. 2012 – http://www.nrel.gov/docs/fy13osti/56806.pdf)
3. *Western Wind and Solar Integration Study - Phase 2* (integration costs only) (Lew et al. 2013 – http://www.nrel.gov/electricity/transmission/western_wind.html)

Analysis Tools

1. NREL's System Advisor Model - advanced tool for estimating levelized cost of energy (LCOE) (http://sam.nrel.gov/)

2. NREL's Cost of Renewable Energy Spreadsheet Tool - simplified spreadsheet tool for estimating levelized cost of energy (LCOE) (https://financere.nrel.gov/finance/content/crest-cost-energy-models)
3. NREL's PVWatts - tool for modeling production profiles of solar resources at different geographic locations (http://www.nrel.gov/rredc/pvwatts/about.html)

4.2.2. How Should Solar Capacity Be Valued?

The authors asked the utilities to explain if and how they valued solar capacity in their input assumptions, modeling, and planning. As shown in Figure 12, some ignored it altogether; a few used wind capacity valuation as a benchmark or proxy, and others have studied solar PV capacity value in detail (believing that wind is a weak proxy).

Utilities that have specifically studied and analyzed the value of solar PV capacity on their system tend to assign some sort of capacity value to PV or they indicated plans to analyze and use values in the near future. Several utilities perform their own analytics comparing solar resources in their region directly to their own load profile, rather than relying upon a generic value or using the same capacity value for solar as they do for wind generation. More recent research found that PV plants in the U.S. West have capacity values that range between 52% and 93%, depending on location and sun-tracking capability (Madaeni et al. 2013). This research also compared data- and computationally-intense reliability-based estimation techniques with simpler approximation methods and found that if designed well, simpler methods can provide accurate approximations. The most accurate of these simpler methods is weighted capacity factor of the plant (Madaeni et al. 2013). By valuing both the benefits and challenges of adding PV to their system, these utilities are able to better capture the positive and negative impacts.

The capacity value can be very specific to a particular location. Even in areas with a strong and consistent solar resource, there are concerns because the system summer peak occurs in the evening hours as commuters are returning home for the evening and just as solar facilities without storage are ramping down. Only through analysis of a specific solar technology, location, and system characteristics can a utility identify the most accurate capacity credit that can be attributed to solar on any particular system.

Also, utilities will perform analysis to compare actual solar performance data to actual load data for increased precision. Comparing actual data to a *forecast* weather normalized load curve can provide misleading results, due to the correlations between weather and load, as well as weather and solar. TMY data (which is an average) can be compared to weather normalized load, or actual solar output can be compared to actual load data.

Importantly, only a few interviewees have considered the capacity valuation differences between fixed and tracking solar PV. The advantage of tracking solar PV is that it optimizes the amount of generation during the evening peak as the modules track the setting sun. The inclusion of tracking PV for those solar installations where it is feasible and cost effective can help optimize the output of the solar modules and reduce the overall impact of shifting the peak into the early evening hours.

Appropriate and accurate capacity values are important in the long-term planning context because they will allow for solar generation to meet at least a portion of future demand requirements. Each technology operates differently in real-world situations and can provide distinct costs and values – only by studying the differences between the two can an accurate solar PV valuation be captured.

4.2.3. How Should Solar Resource Variability Be Considered?

Solar resources can vary greatly depending on where they are located. There are a few key factors that impact the amount of electricity that can be produced from a solar resource, including total solar insolation (or total sunshine on an hourly, daily, seasonal, and annual basis), cloud cover (minute-by-minute and hourly changes), shading (seasonal), and technology used. The combination of these factors and others will determine the scale of impact that one or more solar systems will have when interconnected to a utility's electric grid.

There are several ways that a utility can reduce the potential impact on its system's resiliency and reliability resulting from the variance of output from solar systems, particularly PV.

1. ***Geographic Diversity***: Solar PV is a modular technology. When solar projects are scattered over a wide geographic area, the variability of the entire portfolio of solar projects is limited and greater certainty of the output of the solar portfolio can be achieved. It is similar to when an investor diversifies his/her investment portfolio so that any issues with one investment can be moderated by success in other areas.
2. ***Integration cost add-on***: Utilities can also address resource variability by incorporating an integration cost adder into their planning analytics. This practice is done at many utilities and could be considered a common practice today. If this approach is taken, utilities will get the best analytic results if they conduct a system-specific (i.e., based on their own generation and transmission systems and resources) study to better understand this method. For example, detailed analytics could be conducted to quantify the potential need for incremental spinning and non-spinning reserves as PV penetration increases. Additional reserves can help the overall system deal with some of the ramping challenges and the best analysis will be on the utility's own system.

4.2.4. How Can Scenario Planning More Optimally Manage Risk and Uncertainty?

In some respects, today's electric sector is starting to quickly move away from the one for which the rules and regulations were established. Utilities and regulators are focused on a wider range of options to meet increasing load than just a small number of large, central-station generators, with transmission to interconnect these large plants.

Today, there are many more supply and demand options to consider, including:

- Generation technology choices (e.g. wind, solar PV, geothermal, distributed generation)
- Demand options (demand response, energy efficiency)
- Flexibility options (storage, ramping existing generators differently than tradition)
- Co-optimization of transmission and generation simultaneously.

The potential combinations of supply, demand, and flexible technology options can feel overwhelming. Rather than address the full range of options, to save time and money several interviewees noted that they use scenario planning to consider risks and uncertainties. However, scenario planning cannot investigate the full range of technologies available and identify the optimal portfolio; it can only consider the handful of scenarios that are determined by the analysts. Moreover, examining a few optimum scenarios in depth may not allow for a full exploration of risks and uncertainty. While scenario analysis can test each variable individually, it may not allow for the full capture of interactions between variables.

Utilities routinely incorporate some form of risk analysis into their long-term planning activities. One common approach is to look at the impacts of modifying one variable at a time on a portfolio's overall metrics. This can isolate, for example, how the natural gas price forecast can impact generation dispatch, fuel mix, and the PVRR of a given set of resources.

Several utilities perform a more comprehensive and intricate set of risk analyses as part of their IRP/long-term planning procedures. This is often done by correlating key variables and running Monte Carlo simulations. These random-draw computer models can test how multiple variables can together impact both the potential results of capacity expansion decisions and the subsequent PVRR for each portfolio under consideration. Correlation among the variables has been used in the past, primarily between factors like load growth and natural gas prices. In the future, it is expected that other variables like carbon costs, prices of renewable energy credits, installed capacity costs, and other major items could be tied together so that as one is varied, the rest move in a somewhat logical and predictable fashion. These more complex and interrelated risk methods can provide a more robust vision of how something as dynamic as a portfolio of resources can interact.

4.2.5. How Can Software Models More Effectively Incorporate Solar?

All of the interviewees model their electric system in order to support their supply planning decisions. While they all use different models for different purposes, the goal is the same: to better understand the impacts of their current decisions on their future outlook.

According to several interviewees, commercially available models do not keep up with the technology options and innovations that are actually available. All utilities mentioned that the software they use has one or more constraints that make them less than ideal. Requested changes or updates to the software may result in customized "one-off" versions of the model for their particular situation. Many of the utilities interviewed expressed an interest in learning more about their software options. This interest included a better understanding of the full functionality of the software options and what software other utilities are using. However, none seemed ready to try another software option until they fully understood their options, due to the large investment of money and time needed. One potential solution could be having a third-party review and explore the modeling software available today.

Second, a few utilities mentioned they were potentially interested in software upgrades. In order to make the investment in the new software option, one utility mentioned a desire to see a direct comparison of the existing software to the new software – a comparison that explains the full range of new features, the errors in previous versions that were addressed and statistics like improved run time, computing power needed, and other key factors. Potentially the software vendors or a third party could compare the improvements and advantages of the new software more clearly so that a utility can feel more confidence in moving forward with an upgrade.

The ability of commercially available models to incorporate accurate wind and solar profiles is quite limited. To better represent wind and solar, one utility said it is working on stochastically varying the wind and solar inputs in its own modeling. Instead of using a simplified representation (either a typical week each quarter, or even an average wind or solar year), this utility hopes to utilize stochastic methodologies to vary the wind and solar weather patterns.[27] So, in the first year, they see one type of pattern, and in the second, a totally different pattern is used to more closely approximate the variability of wind and solar generation over seasons and years. The authors did not investigate how these more accurate wind and solar inputs could be incorporated directly into the commercial models.

4.2.6. Should Customer-Sited, Net Energy Metered Photovoltaics Be Treated as a Net Load in Modeling?

Historically, utilities have not had to model or plan for a substantive amount of distributed generation. In the last 5–10 years, the number of customer-sited PV systems has increased dramatically, and utilities are struggling with when and how to incorporate customer-sited PV into their planning and modeling. More background on customer-sited DG can be found in Section 3.4.2. Interviewees raised concerns that span several different aspects of customer on-site generation using solar PV.

Solar is often treated by utilities as a net load. The advantage is that the modeling run time of the analysis is greatly reduced. There are several challenges with this simplification. Several utilities note that load assumptions are one of the main drivers of differences between scenarios tested. Therefore if solar is treated as net load, the analyst is unable to capture the direct impact of solar PV on the system at the distributed or bulk level. Variations in load growth include DG penetration variability, but are not explicitly visible.

One utility clarified that because solar PV penetration is still "in the noise," it is easier to model solar as netted out of load. Even when that utility modeled distributed solar PV as a resource, it found the results didn't differ from the baseline model. Only as PV penetration reaches a certain level does it make sense to model PV as a resource, according to this utility. Several utilities did ask questions about the threshold level – when is this point reached and how will they know they are approaching it? One solution could be for a third party to analyze where the point of inflection is: at what point does including distributed solar as a resource impact the system and is that inflection point different depending on the characteristics of the system?

Another utility treats customer-sited generation as a resource and does not net it out of load. In others words, even customer-sited (residential and commercial) generation competes against other generation in the dispatch queue. Similarly, they treat energy efficiency and conservation measures as a resource that can also compete and whatever resource is most cost effective wins.

This utility operates in more than one state, which can complicate the modeling because some PV and conservation measures are more cost effective in some states than in others. Even so, the utility believes that treating DG and EE as resources is a more accurate way to plan for and model these non-traditional resources. By building in DG, EE, and utility-scale generators as resources, this process allows all technologies to compete against each other on the biggest driver – cost. Together, the combination determines how much supply there will be, what resources will be available in the energy profile, and what choices are accessible during the system peak.

If a utility were interested in treating DG as a resource instead of as a net load impact, it could take its DG forecast curve (or curves, if multiple scenarios are possible) and assign it as a zero cost "must take" resource. The CEM would select it each year in lieu of other options. The PVRR of this portfolio could be compared to the PVRR of the base case run (i.e., without

DG) to determine the overall cost savings from this resource. This type of information could be useful should the utility be interested in leveraging the resource planning process for DG incentive structures.

Additionally, another utility suggested including reserves for these "resources." If DG and EE are included as resources, then utilities will likely want to include this when calculating their reserve requirements. While taking this approach does increase modeling run times and expense, one utility feels the trade-off is worth it to the increased accuracy of choosing the best combination of resources to lower overall system and operational costs.

4.2.7. How Should DG Impact on the Distribution System Be Managed?

The impact of DG on the distribution system is uncertain because it has not been fully analyzed by most utilities. Utilities are interested in both the positive and negative impacts of solar on the distribution system. Until solar reaches a high level of penetration on most utility systems, especially at the feeder level, doing this type of detailed analysis is not considered to be warranted.

Most of the participating utilities acknowledged that DG penetration is expected to become notable in the next 10 years; however, a few interviewees said they were not sure how to determine the point at which DG should be included in their supply planning analyses. One utility also said it was not sure about the methodology needed to include DG penetration in its bulk power analysis studies.

New analyses could help define what is needed and when. Since this is an issue that spans across utilities, it might make sense for a national research institute, like a national laboratory, EPRI, or SEPA, to scope out and perform this analysis to inform the utilities about the general conditions for including DG. This broad analysis would allow for a starting point from which each utility can consider its unique situation.

4.2.8. How Should Net Energy Metered DG's Impacts on Revenue Be Considered?

Although the resource planning process and resulting documents are not part of ratemaking procedures, utility supply planning does occur within the broader context of utility decision- making and is impacted by the utility's overarching goals. As such, during the discussions of resource planning, the issue of the impacts of distributed generation on utility revenues and the implications for utility planning processes arose. Here we present only the results of our discussions with utility representatives, and do not holistically explore the issue of utility revenue loss resulting from distributed generation. This key question indicates that further research is needed. For more thorough discussions of the topic, please refer to other reports (Newcomb et al. 2013, Aliff 2013, EEI 2013, Lehr and Binz 2013, and Rábago et al. 2012).

The rapid adoption of distributed generation is a concern for utilities as it represents a paradigm shift within the traditional electric utility business model. This is a hot topic in many forums held across the country and was identified as a critical issue in utility interviews.[28] Utilities may want to consider examining the tipping point to determine when DG should be included in supply modeling, as well as looking at DG as both a net load and as a resource to quantify the difference in revenue required per kWh. With this information, any differences that exist can be discussed as part of a broader long-term planning stakeholder initiative.

Several interviewees mentioned their anxiety about the possibility that their distributed generation incentive programs and/or NEM programs might be more popular than anticipated. In that case, one utility mentioned that the success could lead to an overall rate increase, particularly as the utility sells less power and has fewer customers to cover the cost of the transmission and distribution system.[29] Few utilities would be interested in creating incentives that result in a rate increase; utilities do not want to fundamentally change the way the revenue forecast is developed. As directed by state laws and regulations, utilities strive to be careful when developing their revenue requirements, so that their attention is on least-cost planning. This least-cost planning focus is especially important for any analyst that considers the market from the wholesale power provider perspective. Their main question is what happens to the rate revenues if this new program is successful?

None of the interviewees identified a direct solution to the possibility of rate increases that result from successful customer-sited on-site generation programs. This is likely because regulatory approval for any potential solution is required and could be challenging.

Considering alternative rate mechanisms may also create economic, social, and political angst within the traditional rate setting process.

Three examples of alternative rate mechanisms are currently under discussion or at the early stages of deployment[30] in the United States, (and more potential solutions are expected to develop in this rapidly emerging field):

1. ***Fixed cost recovery via demand charges, customer charges, or standby charges***
   In this approach, the utility modifies their residential rate schedules to recover some fixed costs through either the creation of a demand charge (which is currently very rare on a residential customer level) or by implementing a fixed customer or standby charge for customers who have DG behind their meter (Newcomb et al. 2013). This approach can be difficult to implement, as stakeholders are likely to see it as a direct charge for "going solar." As one utility pointed out, new technologies may require new ways of thinking on the rates side as well.
2. ***Create a "Value of Solar" rate schedule***
   A second approach used in Austin, Texas actually creates two separate transactions (Rábago et al. 2012). The customer continues to pay all applicable charges on their bill to the utility. Under a separate transaction (combined on the bill for simplicity), the utility purchases all of the PV system's output from the customer at a rate that represents the value provided to the utility of that resource. This approach is beginning to gain traction. It will require a sound methodology on how to calculate the value of the resource, how to translate that into a contract rate, and what the update frequency of that rate will be. Lessons learned are starting to emerge on the program design, which others may want to consider. One utility mentioned that this is very complex and confusing for all parties (particularly customers), and that it is challenging to get consensus on appropriate methodology and the values to use.
3. ***Reframe the cost of service framework***
   The final method is to develop and implement correct market and regulatory incentives to reframe the cost of service framework (Lehr and Binz 2013). One way is to perform a "line item" cost review of all of the services that the utility provides to the customer, and likewise the services that the DG customer provides back to the

utility. These two buildups can then be netted for purposes of customer billing. While technically sound, this is a very complex approach.

4.2.9. Can Solar Supply Planning Be More Appropriately Linked to Other Utility Planning and Operations?

Long-term supply planning performed at utilities does not always impact supply procurement or transmission planning. For solar generation resources, particularly at the distributed level, this may not allow for the full value of these non-conventional supply sources to be captured by the utility unless solar supply planning is linked to other utility planning and operations. This section explores two of these factors.

Several utilities interviewed acknowledged that the tasks of analysis for generation supply planning and issuing RFPs in order to procure new supply fall to different groups. These utilities indicated that the two groups may not extensively collaborate or coordinate. Without more coordination, the RFP evaluation criteria may not reflect the future needs or expectations of the utility adequately, and future supply planning could then have to be adjusted to accommodate different generation resources than were originally modeled and anticipated as optimal.

Several utilities discussed using their internal bid database from previous RFPs as a benchmark for their anticipated RFP price results ("mark-to-market"), which is standard practice.

One utility mentioned that in addition to examining individual RFP responses and comparing them on a price basis alone, the short-list of potential awardees could be modeled in the current system to determine which of the bids actually delivers the best value to the existing portfolio, beyond just the price proposed. This is a tested approach used by other utilities (Mills and Wiser 2012). This type of detailed modeling allows more factors than least cost to be considered, including impacts on the local system, resource diversity, and supply resources that lower overall system risk. While the utility hasn't yet implemented this type of approach, it is considering it (not only for renewables, but for all supply resources).

Another option that goes beyond the mark-to-market test is that the supply planners can be more directly involved in developing the RFP evaluation criteria. If there are specific characteristics that are desirable in the optimal supply portfolio, these can be communicated to the supply acquisition team; the supply planners also may be able to help develop specific criteria associated with each key factor. By including these criteria in the RFP and clearly articulating what the utility needs to best meet the optimal supply future (e.g., supply located in specific locations, specific technologies), the chances of getting new supply that meets those needs increases dramatically.

4.2.10. Can Transmission and Distribution Planning Be More Appropriately Linked to Supply Planning?

For most of the interviewees, supply planning does not usually inform transmission planning analysis or decisions.[31] The two are regularly performed separately and do not always inform each other. While this is standard practice at the utilities interviewed, the increase in penetration of renewables, and particularly solar at the distributed level, may mean that there are unique costs and benefits that could be considered. In these cases, it might be advantageous for the generation supply and transmission planning teams to more closely coordinate. Without this discussion or additional analysis, the value of T&D deferral benefits that result from DG is likely assumed to be zero.

One utility interviewee already links generation and transmission analyses. The utility integrates generation and transmission planning through a more complicated analysis, but the utility staff feels they can identify the optimal combination of supply and transmission by examining many more scenarios. The interviewee considers many different scenarios that can dramatically change the generation supply portfolio (along with associated constraints – e.g., range of natural gas prices, CO_2 scenario). The utility also identifies a handful of new transmission build-outs across its system footprint. Then it considers the full combination of generation supply under each transmission build-out scenario. The utility analyzes the full range of combined supply and transmission portfolios to satisfy all of its stakeholders in a transparent, collaborative way.

Another utility knows that it wants to better link supply and transmission planning and has started internal discussions toward that end. The supply planners are talking to the transmission planners to think about whether localized penetrations of solar will change the delivery point load forecast in their service territory – and to consider how it could result in changes or impacts to the transmission plan. This utility hasn't yet chosen a path forward.

4.2.11. How Can a Utility Consider Solar During Periods of Low/No Load Growth?

Today, many utilities have more generating capacity than they need due to the economic downturn. The combination of less manufacturing and industrial activity and cost-cutting measures, as well as improved energy efficiency and conservation, has flattened out the overall demand for electricity more quickly than anticipated. As a result, some generation and capacity resources added in the past decade may not be fully utilized.

Utilities continue to study new supply technology options, despite low or no load growth, for two main reasons. Many utilities are studying the impacts of repairing/retrofitting existing plants instead of replacing them. One example is whether to retire a coal plant and build a new replacement facility, retrofit it with environmental controls, or to replace the generators with alternative technologies. New supply options could inform which technologies are used for replacement.

Second, several utilities pointed out the current lack of demand growth is a near-term issue for all resources, not just renewables. Sometime in the future, new supply planning will be needed. One utility suggested the desire to investigate and analyze the full range of technology options available today, taking time to better understand those options. In so doing, the utility suggested that full range of costs and values could be considered, including ways to reduce overall portfolio risk (e.g., increased diversity), ways to hedge against or reduce uncertainty (e.g., potential future fuel price scenarios), and better understanding of the cost of the full range of alternatives.

4.2.12. Other Considerations for Solar in Capacity Expansion Models and Ongoing Research

Modeling an electric system can be challenging for sub-hourly timing specifically and for integration modeling more generally. This section examines the challenges with sub-hourly modeling, capacity expansion models, and explores the impacts of broader modeling approaches, based on current NREL research. Wind and solar can vary on a minute-by-minute basis, whereas most production cost models run only hourly simulations over a year. Therefore it is challenging for today's production cost models to run sub-hourly and to consider how to integrate the wind and solar resource more closely to actual production. Likewise, running detailed cost models at 5-minute levels can drastically increase run times and may not be practical overall.

Some utility analysts have performed analysis of integration costs at 5- or 10-minute intervals, but not many. These utility analysts estimate integration costs in the range of \$2–\$11/MWh, as explained in Section 4.4.1 (Mills and Wiser 2012). Importantly, this range reflects assumptions used by utilities, not the range of estimates found in detailed studies of PV variability (due to the paucity of actual studies). Utilities can examine these previous studies and use the information found therein as proxies for the cost of solar integration on their system, as a first-cut approximation.

Another option is to run the analysis over shorter timeframes. If the utility's goal is to determine solar feasibility, then detailed simulations of "tough" periods make sense. Instead of running a 5- or 10-minute model for the entire year, one utility suggested that running "one-off" cases can help, by examining a month or two at 5- or 10- minute intervals. Another utility suggested that modeling could be performed using a 5-minute dispatch on a representative day or week each month during the planning horizon. Either way, utilities could be able to study variability impacts without having to wait for the entire year to run.

If the utility's goal is to improve the accuracy of PVRR estimates, then the utility might only need to run an hourly model with additional operating reserves to represent the cost of resources that will be needed to chase sub-hourly deviations.

Several of the methods and assumptions used to characterize renewable energy technologies in CEMs can significantly impact the simulated value of solar generators, and associated deployment of solar resources in utility IRPs. Ongoing research at NREL and within the academic community has identified some modeling choices that are likely to overestimate the value of solar, while others under-represent the contribution of solar to system performance. The magnitudes of these impacts are frequently utility- and scenario-specific, and this is an area of active research to identify modeling methodologies that can accurately capture the value of solar generators to system performance.

Based on current knowledge, four common modeling approaches are discussed below, along with their impact on simulated solar resource selection.

- **Addition of solar capacity to previously-optimized generation portfolios can lead to suboptimal resource selection.** Some utilities optimize capacity expansion without renewable resources, and then add solar resources to the conventional generation fleet without accounting for the energy and capacity contributions from solar generators. These portfolios do not represent a least-cost resource plan because they do not co-optimize renewable and conventional generation resources in the CEM (Mills and Wiser 2012). This practice is akin to assigning solar a 0% capacity value, and results in higher simulated costs of adding solar resources to a utility portfolio than if the CEM had co-optimized the deployment of conventional and renewable generation resources.

Table 5. Challenges and Utility-Identified Best-Practices for Incorporating Solar Into Utility Supply Planning, Based on Utility Interviews

Main Challenge	Challenge Details	Potential Solutions
Future Solar PV Price Uncertainty	What is a credible estimate of future solar PV prices? Or a range of predictions?	Gather and review reputable data sources, many of which are publicly available. Compare a wide variety of public historical costs and predictions of solar (and other renewable) costs.
Solar PV Technology Characteristics	How should I value solar capacity, if at all?	Utilities that have studied the value of solar capacity tend to assign some value through solar-specific analysis. Study the capacity value of solar specific to your location, technology and utility system. Consider tracking, as it may reduce the system peak shift into the early evening.
	How concerned should I be that solar PV output can vary greatly?	Dispersal of solar PV over a wide geographic area can diversify output and create overall greater certainty in total solar output.
Risk and Uncertainty – Limitations of Scenario Planning	Scenario planning may not address full range of risk, uncertainty, and the full range of interactions between the supply, demand and flexibility technology options available today.	Develop more comprehensive and more inclusive risk analysis to address today's complex electricity market and technology options.
Modeling	Commercially available models don't keep up with or include technology innovations; software is sometimes less than ideal.	Explore and compare software options – it would be particularly helpful for a third party to compare the software available today (to available packages as well as upgrades). Optimize wind and solar modeling inputs and move away from simplifications.
	Integration at the sub-hourly level is challenging using today's models.	Test 5-minute intervals for 1 or 2 key months instead of the entire analysis period. To improve PVRR, include additional operating reserves in hourly modeling.
Customer-Sited PV	Load assumptions drive scenario results.	Treat both DG and EE as resources instead of as net load
	Uncertainty of how DG impacts the distribution system.	Perform analysis to understand when and how DG needs to be included (perhaps through a third party). Begin to model on-site generation as its own resource to see impact on distribution.
	Desire to avoid a rate increase.	Consider alternative rate mechanisms, including solar tariff rates (utility collects for services and self-generators get payment for their full benefits).
Link Solar Supply Planning to Other Utility Planning and Operations	Resource RFPs are not always linked to supply planning.	Go beyond mark-to-market solar database; consider modeling RFP finalists to understand their value in today's system; also coordinate with supply planners to include specific RFP criteria in procurement.
	Transmission planning and decisions are not always linked to supply planning.	Simultaneously examine optimal generation and transmission planning through combined analysis.
Low/No Load Growth	Load growth has slowed or halted and utilities have more generating capacity than they need.	Analyze the full suite of potential supply and demand technologies available for replacing retiring plants, compared to environmental control retrofits. Better understand options to meet future resource needs.

- **Failing to account for decreasing marginal capacity credit can inflate the perceived value of solar generation to a utility system.** As outlined in Section 3, the marginal or incremental capacity value of new solar installations can decrease as more solar resources are installed, because the utility's peak in net load shifts from afternoon to evening. Failing to account for this by dynamically calculating the capacity credit of solar generation as a function of installed solar capacity will likely result in CEM scenarios where solar is expected to provide more firm peak capacity than is feasible. This can over-estimate the value of solar generation in a utility system and lead to higher than optimal solar deployment.
- **Renewable generation and other infrastructure investments by other, neighboring utilities can impact optimal solar investment in a utility's own service territory.** Utilities in the United States operate within synchronized electrical interconnections where new generation resources outside a utility's service territory can impact transmission line congestion and the price and accessibility of power transfers from other regions. Several recent studies have analyzed renewable resource build-out pathways and their impacts on power flow and marginal energy costs (Mai et al. 2013; Milligan et al. 2012; Lew et al. 2010). These studies suggest that failing to account for out-of-state capacity expansion either directly or through dynamic boundary conditions can lead to sub-optimal investment within a utility's jurisdiction. For example, if high solar deployment is expected in a neighboring region, a utility might cost-effectively choose to rely on market purchases during hours of peak solar generation when solar electricity may suppress marginal electricity costs in boundary regions, rather than developing solar resources within their service territory. The sign and magnitude of this impact is highly system- and scenario-specific, and could lead to over- or under-estimates in simulated optimal solar deployment.
- **The representation of thermal generator operating limits can impact the optimal level of solar resource investment.** To estimate production costs and determine the most cost-effective combination of generating resources to meet load, some CEMs rely on a load duration curve-based approach, and others use a simplified, chronological dispatch. The latter approach can be improved by modeling generator commitment states as integer or binary variables, rather than continuous variables, in order to capture important generator characteristics such as minimum stable generation level and minimum up and down time. A chronological approach with explicit representation of generator commitment status has been shown to more accurately model the impacts of solar production on the cycling and ramping behavior of thermal (gas and coal) plants (Palmintier and Webster 2011). CEMs that do not model these constraints are likely to find sub-optimal generation portfolios and fail to represent the true impact of solar production variability on system cost, which could lead to over or underestimations.

These and other modeling considerations are likely to have important impacts on the selection of solar resources in utility IRP processes. Building on prior analysis (Mai et al. 2013), current work at NREL aims to more accurately quantify the impacts of these and other modeling approaches and assumptions on optimal solar deployment levels in utility planning processes (Mai et al. 2014 Forthcoming). This research is focused on identifying ways that commercial CEMs could be improved to more accurately represent renewable energy value within a utility portfolio. Specifically, near term research will use a newly developed CEM, NREL's Resource Planning Model, to quantify the sensitivity of optimal solar deployment to different methods for quantifying solar capacity value, different representations of renewable development in neighboring utilities' jurisdictions, and different methods for modeling thermal generation and transmission constraints. Preliminary findings using Colorado-area utilities as an example are expected to be published in early 2014.[32]

4.2.13. Summary of Solar Supply Planning Challenges and Solutions

Table 5 summarizes the challenges utility interviewees identified for incorporating solar PV into utility supply planning, as well as some potential solutions. While this is not comprehensive, this information may advance the discussion of how to better represent solar energy in utility supply planning.

5. Utility-Identified Analysis Needs

Solar generation is becoming a viable resource for inclusion in IRP/long-term planning analytics across the country. In some regions solar generation already competes from a cost perspective, while in other states and regions, RPS requirements may still drive solar adoption. Through utility discussions, the authors identified several areas that warrant further consideration and research to better understand the impact of solar on local and regional utility operations, and how it can be optimized to best support the resiliency of the electric grid. These include:

- ***Credible PV price and performance data*** – Identify and gather PV price and performance data. Utilities could consider centralizing the data into a database with a trusted third-party, who could aggregate the information so the breadth and depth of the data can be shared between utilities, without specific project or utility identifiers.
- ***Analysis of how to incorporate geographically diversified resources into modeling*** – Utilities today typically allow their CEM to consider a standard size (MW) for each potential resource it may select (i.e., a 50-MW solar generator, or a 110-MW combustion turbine).

  Solar is unique in that its footprint is scalable to any size requirement and can be integrated across a large geographic region. Rather than installing a static 50-MW solar farm at one location, a utility could choose instead a set of ten 5-MW solar farms spread out across its service territory. This approach is not typically used in IRP processes today, but further analysis could explore the possibilities and explain the modeling implications. Question for investigation could include – what size makes sense for these smaller farms? Where could they be realistically located? And how does spreading them out impact local and overall system operation?

  These smaller systems would be easier to interconnect (due to the fact that they could tie in at lower voltages), and the variable and intra-minute fluctuations (i.e., quick ramping) due to cloud cover and other weather events could be smoothed. Staggering systems from east to west in a service territory could stagger ramping up and down across the aggregate production. Such an approach, while not currently implemented in IRPs today, may be viable and provide a glimpse into how solar can be optimized in the future. Additional analysis to explore different configurations of smaller solar farms could be informative to utility decision-makers.

- ***Analysis of the potential relationship between energy storage and PV*** – Energy storage applications are an intriguing match for solar generation. Although storage is less important at current low penetrations, at high penetrations storage could be useful in smoothing fluctuations in solar production, or as a method to provide firm output across the system peak. Storage may be particularly helpful for distributed applications, to help address potential islanding due to high percentages of distributed solar on a particular feeder. There are several pilot projects that tie solar and energy storage together, but to date there has not been long-term analysis of efforts to bundle these technologies.

  The primary roadblocks to incorporating energy storage are its high cost to deploy and uncertainty in how to value the resources it can provide. Unlike solar generation (or any fossil-fuel resource), energy storage does not actually generate power; rather, it stores it, presumably from a less expensive time to a more expensive (peak) time. This arbitrage by itself is difficult to cost justify because of the inherent losses in storing energy. Pairing storage with solar generation to increase its capacity value and providing both smoothing and firming services may create sufficient justification for deploying energy storage resources.

  Current analysis explores some key aspects of energy storage including operational benefits (Denholm et al. 2013), market and policy barriers (Sioshani et al. 2012), high penetration of renewables in combination with storage (Augustine et al. 2012 and Denholm and Hand 2011), and ways that storage can reduce renewable energy curtailment (Denholm 2012). More analysis on the benefits and different configurations of solar and energy storage could inform ways to consider specific storage-PV combinations, and their direct impact on utility operations.

- ***Easier ways to predict impacts of increased PV penetration*** – In Figure 13, varying levels of PV penetration were shown with their relative impact on a utility's peak load day. The higher the penetration, the less its incremental capacity will be able to contribute to meeting the utility's peak demand.[33] Currently, CEM software tools appear unable to model this dynamic impact to capacity values.

  High penetration of PV and renewables has been analyzed, including large-scale solar deployment (Drury et al. 2012) and technologies that enable high wind and solar penetration (Denholm 2011). Including these capabilities and modeling the interactions more closely could allow for a more robust picture of system interactions at higher levels of solar penetration.

- ***Better risk/uncertainty analysis methods (beyond scenario planning)*** – Examine and consider performing more comprehensive risk analyses as part of long-term planning, because scenario planning is not capable of investigating the full range of technologies available in order to identify the optimal portfolio. Using an advanced risk analysis method (e.g. Monte Carlo), correlate key variables (load growth, natural gas prices, carbon costs, renewable energy technology prices, installed capacity costs, etc.) so that if one is varies, the rest move in a logical fashion.

- ***Improved commercial production cost models*** –Include technologies in analysis that are not currently available or poorly represented in commercially available models. Today's PCMs are only capable of doing a resource dispatch hourly. However, solar output may fluctuate enough within the hour from cloud cover and other weather events to create operational impacts that are not currently captured. Several utilities mentioned a desire to be able to model the variability of solar (and other variable renewables such as wind) on an intra-hour basis.
  Intra-hour cost modeling is not currently done because running PCMs on any time scale less than an hour would require massive amounts of computing power, or slow the model run time such that analysis would take significantly longer timeframes. As explained in Section 4.2.6, this timeframe could be run for a few key months, or for a few representative weeks each month. This level of analysis may be sufficient to reveal the implications of solar production variances and the level of flexible quick-start generators required. In addition, improvements in PCMs themselves (i.e. decreased run-times) or using faster, more powerful computers (i.e., high performance computing centers) could address these issues so intra-hour solar could be modeled more easily.
- ***Translate distribution system impacts to long-term plans*** – Much discussion has occurred in recent years regarding the true value and cost of solar generation, particularly as it relates to the distribution grid. Solar advocates believe that system upgrades can be deferred and asset lives extended due to solar being located on local feeders. Utilities are more skeptical and see the potential need for significant system upgrades in the future because the existing distribution system was not designed for bi-directional flow. While IRPs do not often address long-term distribution system planning, effects on distribution system values and costs could become important for incorporation, particularly as NEM expands across the country.
- ***Clarity about when to include distributed generation in supply modeling*** – PV systems are often tied in behind a customer's meter. As such, these systems can be modeled in an IRP as a net load impact that highlights reduced retail sales in the future, or as a stand-alone resource. The method used impacts the PVRR on a per kWh basis for all customers. Utilities may want to consider examining the tipping point to determine when DG should be included in supply modeling, as well as looking at DG as both a net load and as a resource to quantify the difference in revenue required per kWh. With this information, any differences that exist can be discussed as part of a broader long-term planning stakeholder initiative.

GLOSSARY

avoided cost The incremental costs of electric energy or other services, if a utility did not purchase from the existing power seller; the focus is on the cost of alternatives available to the buyer/utility.[34]

azimuth The angle between the north direction and the projection of the surface normal into the horizontal plane; measured clockwise from north.[35] More simply stated it is the angle between due north (0 degrees) and

	the direction from which the sunlight is coming. On the equinoxes, the sun rises at due east with an azimuth of 90 degrees, and sets at due west with an azimuth of 270 degrees. At solar noon in the northern hemisphere, the azimuth angle is 180 degrees.
baseload generation	Electric generating facilities within a utility system that are operated to the greatest extent possible to maximize system mechanical and thermal efficiency and minimize system operating costs.[36] Baseload generation typically has annual load factors that exceed 75%. Examples include coal-fired, natural gas combined cycle, nuclear, very large, damned hydroelectric (i.e. not run of river), geothermal and biomass.[37]
capacity expansion model (CEM)	A computer software tool used in resource planning to determine potential expansion of electricity generation, storage and transmission systems over several decades. The model chooses the cost-optimal mix of technologies that meet reserve requirements, technology-specific resource constraints and policy constraints.[38] Results are used to generate plans for electricity capacity additions or energy purchases, and are tested in production simulation models to determine the final cost and emissions outputs. Inputs to the model include variables such as existing generating assets and purchased power contracts, potential resources, load growth assumptions, fuel cost curves, capital and operations/maintenance expenses.
capacity factor (CF)	The ratio of actual energy produced by an energy generating unit or system in a given period, to the hypothetical maximum possible (i.e. energy produced from continuous operation at full rated power).[39]
capacity value	The value assigned to a generating facility, based on the extent to which the facility can help reliably serve load.[40] In this report, we use capacity value in the sense used by resource planners (expressed as a percentage of maximum generating capacity), rather than the monetary or market value assigned to the capacity.
derate factor (DC/AC)	A number accounting for the loss of power in the conversion from DC to AC power. The overall derate factor for a system is the product of the derate factors for the components of the system.
distributed generation (DG)	Generation located on a utility distribution system, typically at or near the load. For purposes of this report, DG is often intended to mean customer-sited PV.
Effective Load-Carrying Capability (ELCC)	The ability of a power generator to support additional peak load without reducing the reliability of the electrical system (in terms of loss of load probability or loss of load expectation). The amount of ELCC for a particular generator is calculated by determining the amount of existing supply capacity that can displaced by the source while serving the same load profile and maintaining the reliability of the electrical system.
energy storage	Devices or technologies that can store electrical energy so that it can be used to meet demand at a later time.[41] Adding an energy storage system to a solar photovoltaic system allows the solar energy to be used when the sun is not shining. This expands the flexibility of the photovoltaic system.

greenhouse gases (GHG)	Gases that trap heat in the atmosphere.[42] Specifically, they absorb and emit infrared radiation and contribute, either directly or indirectly, to the process of absorption and re-dispersion of thermal radiation coming from a planet's surface (the greenhouse effect). Greenhouse gases include water vapor, carbon dioxide, methane, nitrous oxide, ozone and chlorofluorocarbons (CFCs).
grid parity	The point where the cost of alternative energy sources equals the cost of electricity purchased from the grid.[43]
heat rate	A common measure of the efficiency of a steam power plant. The amount of energy input to an electric generator to generate one kilowatt hour of electricity generated; typically represented in Btu/kWh.[44]
integrated resource plan (IRP)	A plan developed by an electric power provider, sometimes required by a public regulatory commission or agency, that defines the short and long term capacity additions (supply side) and demand side management programs that it will undertake to meet projected energy demands.[45]
integrated resource planning (or resource planning)	The process of developing an integrated resource plan.
intermediate generation (load-following power plants)	Power-generating equipment that can vary its level of output (generation) in response to changes in electricity demand. Normally operated on a daily cycle to serve on-peak loads during the day but not off-peak loads during nights and weekends.[46] Typically, these plants have annual load factors ranging from 40% - 60%. Examples include smaller coal-fired power plants, natural gas combined cycle plants, most hydroelectric (when the weather cooperates), and in the future, offshore wind power, concentrated solar power, thermal solar power and wave energy.[47]
load duration curve	A graph that illustrates the average (or peak) hourly load, from highest to lowest, sorted in decreasing order for all 8,760 hours per year.[48] The area under the load duration curve represents the total demand for the period of time. Load Duration Curves are used in capacity planning.
Monte Carlo	A problem solving technique used to approximate the probability of certain outcomes by running multiple trial runs, called simulations, using random variables.[49] A class of computational algorithms that use repeated random sampling, by running simulations many times over, in order to compute the probability of an occurrence; especially useful for simulating systems with many variables, or coupled degrees of freedom with significant uncertainty.[50]
net dependable capacity	The maximum amount of electricity that a generating system or facility can reliably produce during the most restrictive seasonal conditions, minus the amount of electricity that is consumed by the facility itself, if any.
net energy metering (NEM)	A policy that allows a customer to receive a financial credit from the utility for power generated by their distributed energy system that is fed onto the electricity grid (net, or after most is used on-site

	by the customer, behind the meter). The credit is used to offset the customer's electricity bill and rules vary by jurisdiction.
peaking generation	Generation from power plants that normally operate to serve loads during peak load times or during system emergencies, and that operate at very low annual load factors (5%-15%). Examples include natural gas combustion turbine, and simple cycle turbine (natural gas- or oil-fueled).[51]
peak demand/peak load	A period of time during the day, month, season or year during which the peak electrical demand/load/use occurs on the electric system. In warm climates, the peak electrical use may occur in the summer, when there is high demand for electricity to run air conditioning. In cold climates, the peak demand may occur during winter when there is high demand for heating.
planning horizon	The period of time (in years) that a planning process covers.
(generation) portfolio	The collection of energy generation assets (power plants or purchased energy) available (or planned) to be used by a utility to meet the load.
present value	The current worth of a cash flow, after considering interest. Future value becomes present value through the process of discounting.[52]
Present Value of Revenue Requirements (PVRR)	A dollar amount that represents the total annual revenue, discounted to present dollars at the time of calculation that a utility must collect from customers to pay all costs and expenses including a reasonable return on investment.[53]
production cost modeling (PCM)	Production cost modeling simulates hourly electric generation dispatch over the entire planning horizon. Production costing models are used extensively in the electric power industry to forecast the expected amount of electricity produced by different power generation units and the expected cost of producing that electricity for a given power generation system.[54] The models can be used to forecast the expected amount of electricity production and to perform sensitivity analyses that account for the expected variation of load over time and the expected availability of generating facilities and, often times, transmission and distribution capacity, while accounting for uncertainties (e.g. fuel price, seasonal variations).
Power Purchase Agreement (PPA)	A legal contract between an electricity generator and a power purchaser for the sale of energy, capacity and/or ancillary services. The contract includes all the terms of the sale and often serves to determine the credit quality of the generating project by financers.[55]
renewable energy certificate	The property rights to the environmental, social, and other non-power qualities of renewable electricity generation. A REC, and its associated attributes and benefits, can be sold separately from the underlying physical electricity associated with a renewable-based generation source.[56]
renewable portfolio standard (RPS)	A legislated mandate for utilities to generate or purchase a portion (generally expressed as a percentage) of electricity sold from certain renewable generating sources, such as wind, solar, geothermal, biomass and biogas.

resource mix	The different types of generating facilities that contribute to meeting the load of an electric system within a defined area.
solar renewable energy credit (SREC)	A Renewable Energy Credit generated by a solar generating system, which indicates the production of a unit of solar energy. SRECs are tradable within some U.S states and regions that have Renewable Portfolio Standards, or voluntary green power markets.
typical meteorological year (TMY)	A set of hourly weather data for a particular location over a period of a year that represent a typical year and are consistent with the long-term averages. TMY files capture typical conditions for your location of interest and are available in TMY2 (most current 30 years) or TMY3 (most current 7, 10, or 15 years) constructs.[57] The data (which include temperature, solar radiation and precipitation) are often used for conducting simulations to facilitate the design and location of solar energy systems.
weather normalization	An adjustment methodology that accounts for weather. Normalization allows for the comparison of energy consumption between different periods of time or geographic locations that have different weather conditions by factoring out aberrations or unusual occurrences (e.g. variables such as higher than or lower than normal outside air temperature).

APPENDIX A – TENNESSEE VALLEY AUTHORITY CASE STUDY

I. Who is TVA?

The Tennessee Valley Authority (TVA) is the largest publicly owned electric utility in the United States. It is a federal corporation that was created in 1933 by Congress as part of the New Deal to help move the country out of the Great Depression.

TVA is the primary electricity provider in Tennessee and also covers areas of Mississippi, Alabama, Georgia, North Carolina, Kentucky, and Virginia. The company works with 155 municipal and cooperative distributors that resell over 80% of the TVA-generated power to their consumers. The remainder of the TVA power is sold directly to 51 large industrial customers and six federal installations. Besides power production, TVA provides navigation, flood control, and land management services for the Tennessee River system.

TVA is a self-regulated utility and is not subject to oversight by any state public service commissions, as are other regulated utilities. This means that state-level policies do not apply to the area within the TVA district and TVA is not required to produce an Integrated Resource Plan (IRP).

The company's 2011 IRP, *TVA's Environmental and Energy Future*, was the result of over 30 months of effort and was the first IRP TVA had completed since 1995. TVA was under no obligation to update the 1995 plan, but by 2008 it was clear that the time had come to reconsider the future. TVA staff redesigned the way the IRP would be developed, driven by a strong desire to improve the methodology for considering and planning for a variety of alternative futures. The resulting methodology and product highlighted here is an example of a thoughtful IRP process with clearly presented results.

II. Envisioning the Future: Developing the TVA Scenarios

As TVA approached the task of producing a new IRP, staff from across the company were brought together to brainstorm the attributes of all possible futures and describe these future worlds. Once the most likely futures had been agreed upon, they were presented to stakeholders using storyboards. After stakeholder vetting, the plausible futures were passed on to planning modelers for transformation into individual scenarios through the identification of appropriate assumptions and constraints. This initial phase of conceptualizing the scenarios and modeling the parameters to match those "futures" was a time-consuming step. In fact, it was so critical that it is now incorporated into TVA's annual planning process.

For their 2011 IRP, TVA developed eight future scenarios:

- Dramatic Economic Recovery
- Environmental Focus as a National Priority
- Prolonged Economic Malaise
- Game-Changing Technology
- Focus on Energy Independence
- Carbon Regulation Creates Economic Downturn
- Reference Case: Spring 2010
- Revised Reference Case: Great Recession Impacts Recovery

These scenarios were based on distinct visions of the future that were vetted in collaboration with stakeholders. For each scenario, uncertainty values were identified (e.g. commodity prices, environmental regulations, etc.) and then used as attributes to describe each future. Forecasts for each attribute were developed, discussed with internal and external stakeholders, and compared across scenarios to ensure reasonableness. A review of the scenario definitions after completion of the study made it apparent that some of the scenarios were similar in many respects; that is, some attribute values did not show significant variations across the scenarios. Over time, TVA has realized the importance of ensuring that each imagined future is both distinct and credible. Making sure that each scenario is sufficiently different and likely to occur avoids wasting time and resources on duplicative efforts and far-fetched notions. For example, there is little need to consider a future with 2% load growth if there is little expectation that this future will occur.

In their upcoming IRP,[58] TVA plans to reduce the number of scenarios, perhaps by half. To accomplish this goal, the 2015 IRP will focus on more distinct futures based on key uncertainties of particular interest to stakeholders. This approach aims to satisfy specific cases of interest to a wide variety of stakeholders (e.g., assumptions regarding penetration of distributed generating resources). TVA anticipates that stakeholder involvement in the creation of these scenarios will improve understanding and buy-in during the more complex modeling phase.

III. Developing the TVA Strategies

Once scenarios were finalized, strategies that represented potential business decisions and portfolio choices that TVA could control were developed. First, key components of the strategies were defined, such as the types of generation technologies from which TVA could select and the development of new transmission infrastructure. Company management and stakeholders were consulted during this phase, to identify what business strategies they were willing to consider. If management were not on-board with a particular component, such as the accelerated development of nuclear, there would be no need to consider it.

Next, development of the strategies was accomplished by combining variations of the key components. A total of nine key components (or attributes) were selected, with input from public scoping efforts. These key components included four model inputs and four model constraints:

- The level of the energy efficiency and demand response included in each strategy
- The amount of renewable resources input in each strategy
- The schedule for coal-fired idling to be tested
- The option of including a pumped-storage unit
- Constraints on the addition of new nuclear capacity
- Constraints on technology and timing of new coal facilities
- Constraints on gas-fired unit expansions
- Constraints on the type and level of transmission infrastructure to support the resource options in each strategy.

Through the consideration of these key components, five business strategies, represented by different resource portfolios for the future, were developed:

1. Limited change in current resource portfolio
2. Baseline plan resource portfolio
3. Diversity-focused resource portfolio
4. Nuclear-focused resource portfolio
5. Energy efficiency/demand response-focused resource portfolio.

The strategies were differentiated from each other by the extent to which each key component was included in the strategy. That is, each key component was assigned a numerical boundary or range for each strategy (e.g., the Energy Efficiency and Renewable Focused strategy included the addition of 5,100 MW of renewables by 2020). In response to stakeholder comments, the target ranges for energy efficiency, demand response, and renewable energy were increased in the strategies during the strategy development process.

The attributes of the five strategies were used as modeling inputs to create optimized generation portfolios. First, an optimized capacity expansion plan was generated using Ventyx's System Optimizer tool, which minimizes the PVRR for each portfolio. The resulting portfolios were then evaluated using an hourly production cost model with stochastics, Ventyx's Strategic Planning Model (MIDAS). The model determined the range of plan costs based on Monte Carlo modeling of thirteen key variables that account for commodity prices, financial variables, operating costs, dispatching costs, and load forecast uncertainties. These variables differ from the key attributes (described above) in that the variables are not business decisions, but variables that are not in the utility's control.

The total PVRR for each resource plan was calculated by using the random set of uncertainty variables drawn for stochastic iteration. The mean value across the plan cost distribution from the set of iterations was used as the reported plan cost of each strategy in a particular scenario.

One portfolio was generated for each of five planning strategies associated with each of the six future scenarios during the draft phase of the IRP study. In addition, there was a portfolio for each of the five strategies associated with the Reference Case. This resulted in a total of 35 portfolios. Each portfolio represented a 20-year plan for capacity expansion.

IV. Selecting Preferred Strategies

To evaluate the 35 portfolios, TVA (with input from stakeholders) created a scorecard. The goal was to identify trends or common characteristics that could lead to the most desirable outcomes.[59] The scorecard organized and communicated the evaluation process, which involved scoring each of the five strategies on its cost, risk, environmental impacts, and other strategic metrics.[60] The scoring process was subject to public comment, during which there were discussions of the trade-offs, constraints and compromises. These discussions, as well as a sensitivity analysis to identify top performing strategies were used to refine the evaluation results during the final phase of the IRP.

The results of this scoring process, as well as a narrative explaining the technology investments associated with each strategy, were included in a draft IRP study report. The report was released for public comment consistent with the requirements of the National Environmental Policy Act (NEPA). A companion Environmental Impact Study (EIS) report was also issued with the draft IRP results. Based on public comment and stakeholder input, adjustments were made to the assumptions and the strategies following the release of the draft findings. During the final phase of the IRP, additional modeling and analysis was completed and led to the development of a Recommended Planning Direction, which consisted of guideline ranges for key components of the resource plan that performed well across most scenarios. This Recommended Planning Direction was submitted to the TVA Board of Directors and was adopted in April 2011.The final IRP defined TVA's short- and long-term strategic direction and identified short-term actions.

V. Involvement and Role of Distribution Companies

In the years leading up to 2008, the 155 distributors of TVA contemplated separating from TVA. As a result, there was not significant involvement of individual distribution companies during the initial stages of the IRP process. However, the association that represents the distributors, The Tennessee Valley Public Power Association (TVPPA), did participate as a stakeholder during the IRP process.

Today, there is more direct collaboration with the 155 distributors, and while they are not closely involved in the IRP, they do have involvement primarily through various committees of TVPPA. There is a wide variety of levels of experience and interest in capacity expansion planning amongst the distributors, with some having little knowledge or interest in the details of new capacity needs or construction. The fuel source and technology by which the electricity they distribute is produced has, to date, not been an expressed as the overriding concern, and distributors are more adamant that the cost of power from TVA remains as low as possible.

The distribution companies are, however, the ultimate implementers of all distributed generation programs (such as distributed solar or energy efficiency and demand response products), and therefore need to understand the benefits and challenges of distributed generation on their system. Likewise, TVA needs to know which distributors will be offering distributed generation programs to their customers and the extent of the customer-interest, in order to estimate the role that distributed generation will play in meeting future load and its effect on future transmission system needs. In some cases, the distribution companies have expressed some reluctance to share information regarding customer involvement in distributed generation with TVA.

The uncertainties associated with the TVA relationship to its distributors, including the potential of separation and the imperfect information regarding future involvement in efficiency, demand response, and distributed generation programs, introduces an element of risk into the TVA capacity planning process. If communication surrounding these uncertainties remains unchanged, this risk will increase with rising penetration of distributed generation on the system. Currently, TVA planners assume an overall level of DG penetration and adoption across the TVA territory; however, a methodology that includes sub-region-specific DG penetration curves may be necessary in the future.

VI. Renewable Energy and Energy Efficiency Program Considerations

For the 2011 IRP, a fixed portfolio for renewable energy and energy efficiency was created outside of the capacity planning model and input as must-take transactions. These portfolios were designed using standard cost-effectiveness tests and other techniques and during the IRP were tested to confirm overall benefit, although individual components of the defined portfolios were not optimized directly in the models. The renewable energy portfolios were pre-selected based on a determination of future markets and the availability of supply. In its upcoming IRP, TVA wants to allow the model to select the renewable energy portfolio from a variety of options based on current market and resource data. This will result in more optimized renewable energy costs and production profiles. A similar approach to energy efficiency resource selection is also planned for the next IRP, based on direct optimization of small blocks of efficiency based on end-use or customer segment groups.

VII. Conclusion

TVA executed a thorough resource planning process that includes significant and meaningful stakeholder involvement, and culminates in a clearly written Integrated Resource Plan. TVA dedicates significant resources to informing, surveying and responding to stakeholders through all phases of the planning process, from scenario formulation to portfolio selection. The TVA IRP explains this process; results are well-organized and clearly written and thus are accessible to all stakeholders.

APPENDIX B – SELECT UTILITY QUESTIONNAIRE RESPONSES

1. **What is your integrated resource planning horizon?**

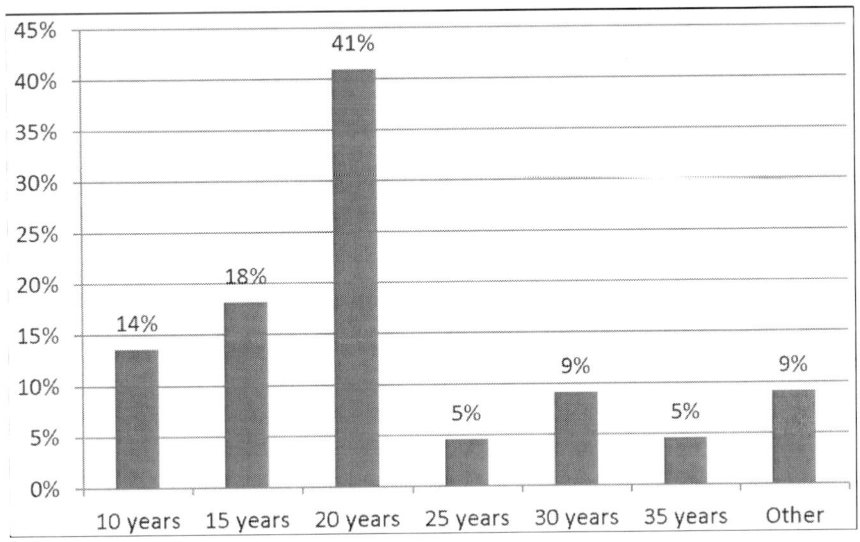

Total number of responses: 22

2. **What is the average annual consumption growth percentage for your planning horizon, net of any energy efficiency or customer-sited generation impacts?**

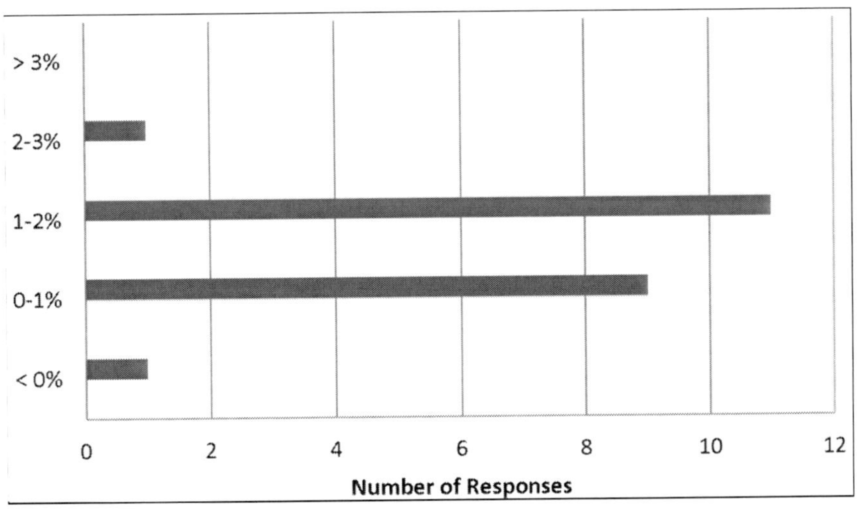

Total number of responses: 22

3. How is distributed, net metered solar (behind the meter) incorporated into the modeling?

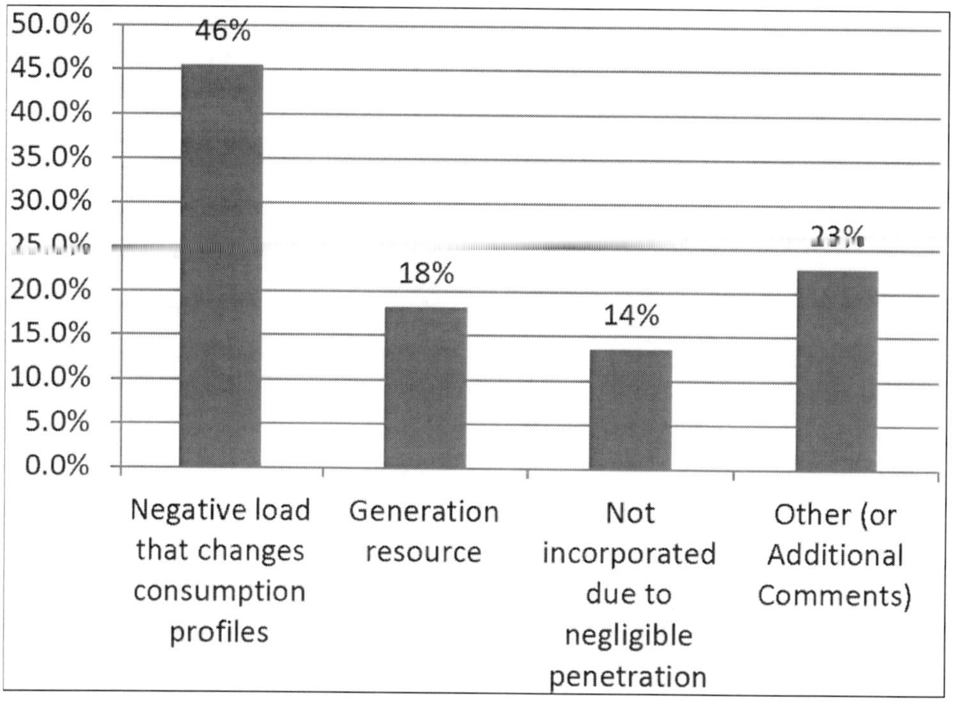

Total number of responses: 22

4. (a) Does your utility incorporate long-term transmission system planning into its resource planning process?

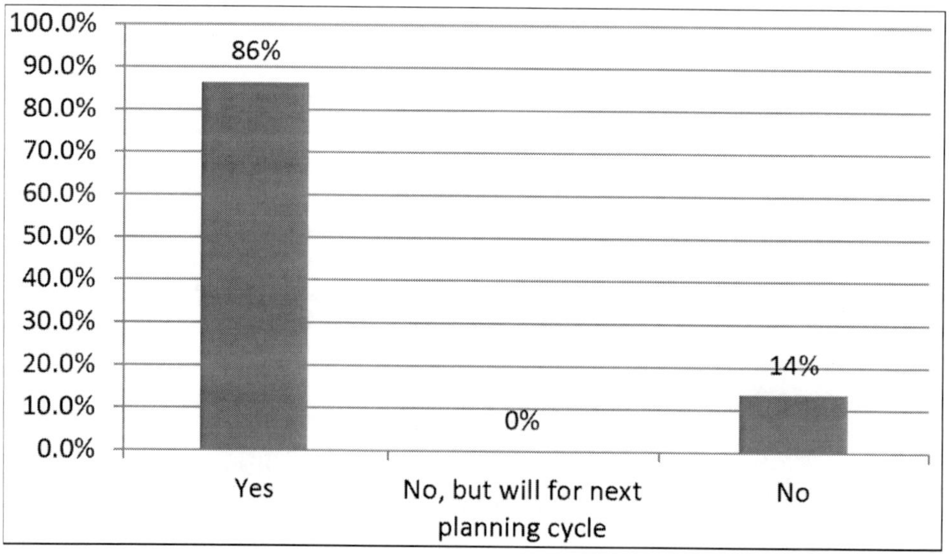

Total number of responses: 22

(b) Does your utility incorporate long-term distribution system planning into its resource planning process?

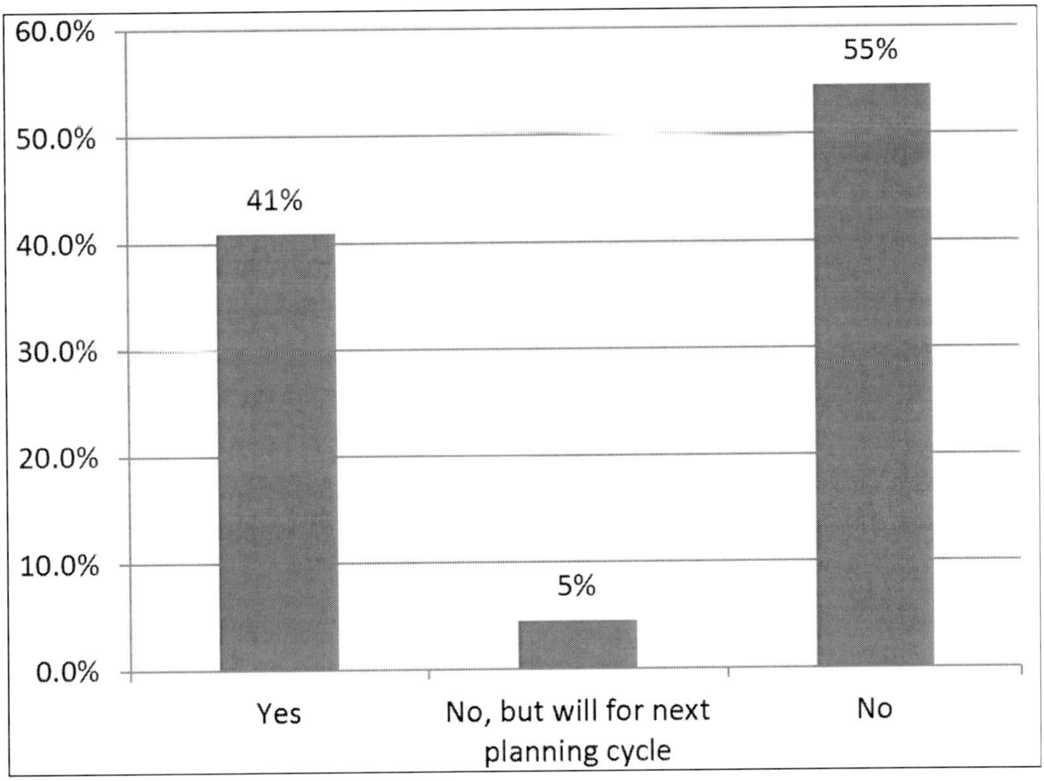

Total number of responses: 22

5. **(a) What capacity expansion model software does your utility utilize? (ex. Strategist)**

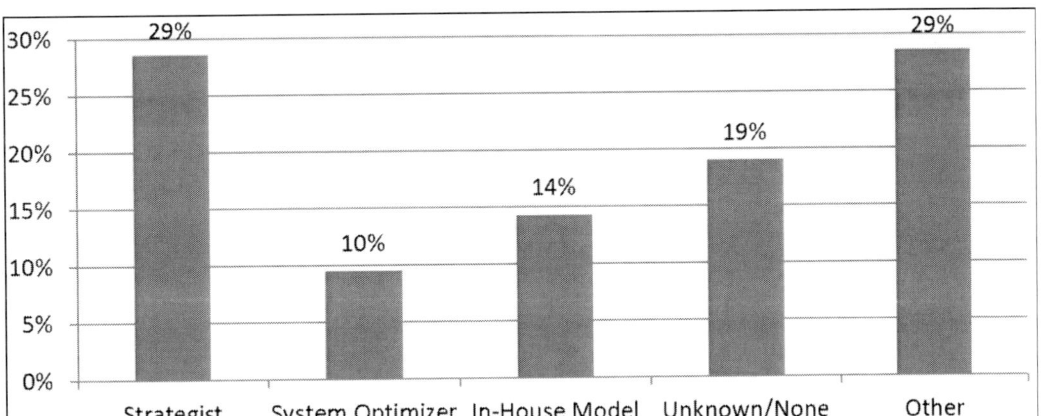

Other: Plexos, Aurora, P&R CapEx, EGEAS, Outsource to 3rd party

Total response count: 21.

(b) What production cost modeling software system does your utility use? (ex. PROMOD)

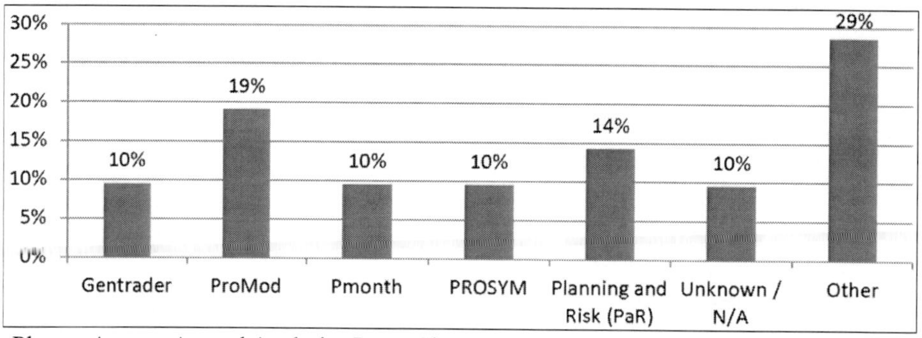

Other: Plexos, Aurora, Ascend Analytics PowerSimm, Market Analytics, RTSim, MIDAS

Total number of responses: 21

6. (a) What type of risk analytics does your company perform?

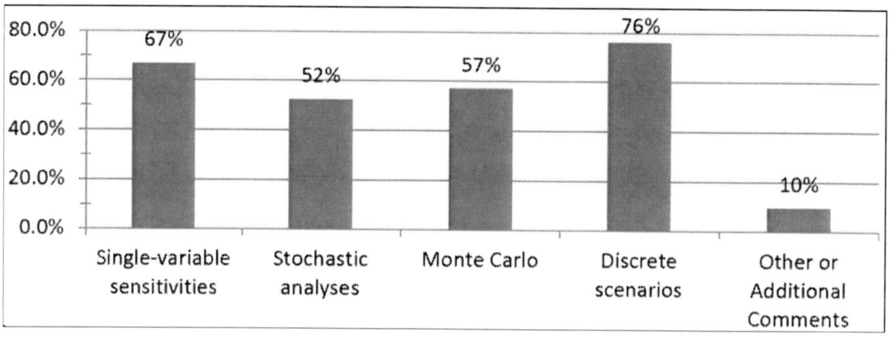

Total number of responses: 21

(b) Which variables do you stress?

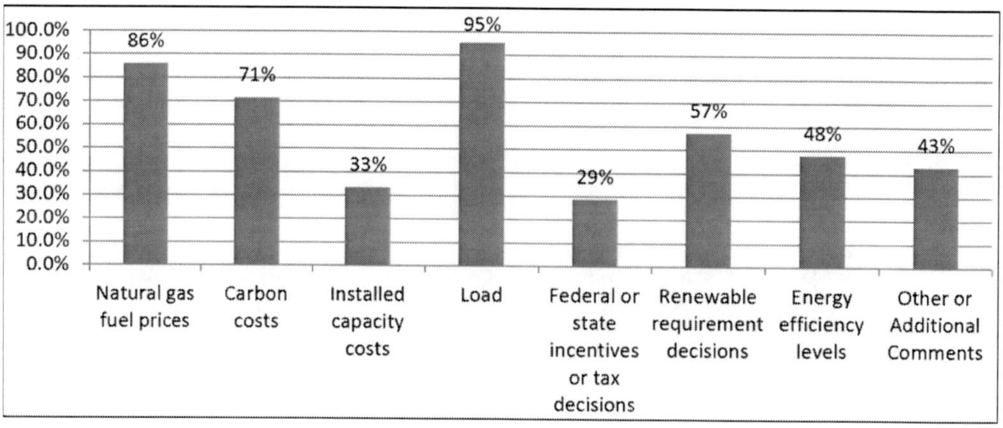

Total number of responses: 21

Other or additional comments:

- We vary hydro output
- Stochastic modeling stresses: load, gas prices, power prices, and outages
- Energy efficiency is a single-variable sensitivity; the other items are stressed stochastically. The other item stressed in the risk analysis is the market price for electricity
- Other commodity prices, financial parameters, O&M costs, availability
- Sensitivities are run around multiple variables.
- fuel consumption, fuel costs, as-available energy purchases, as-available energy curtailment, capital cost for new generation, independent power producer payments/costs
- High/Low Distributed Generation
- EE Costs, Inflation
- Fuel consumption, fuel costs, as-available energy purchases, as-available energy curtailment, capital cost for new generation, independent power producer payments/costs

7. How do you choose solar capacity sizes for candidate portfolio evaluation?

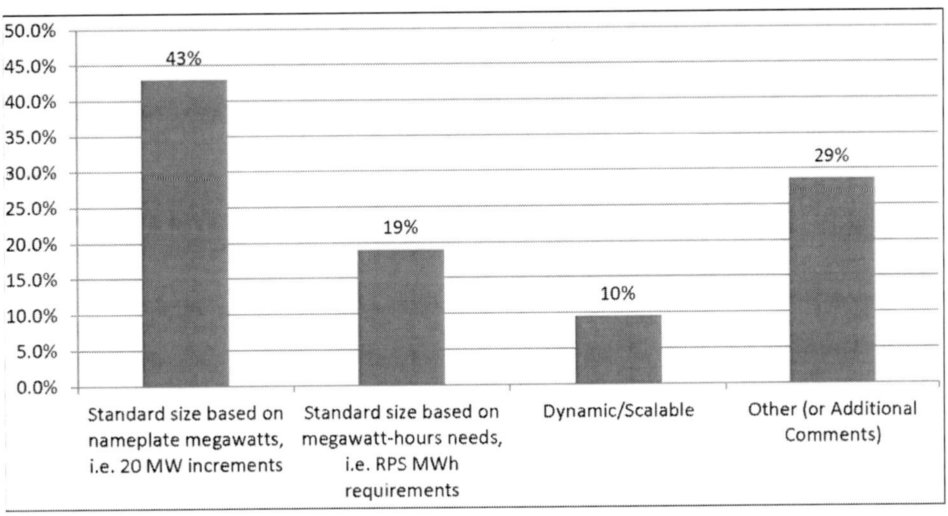

Total number of responses: 21

Other or additional comments:

- Solar capacity size is standard based on MWh needs (i.e., RPS MWh requirements)
- Based on customer request
- Sizes are based upon actual projects
- To meet our Sustainability Targets - 20% by 2020, solar is one of many sustainable options that is valued
- Policy directed from governing board

8. (a) What capacity values for resource adequacy purposes do you assign for each type of solar resource? Note: not capacity factors

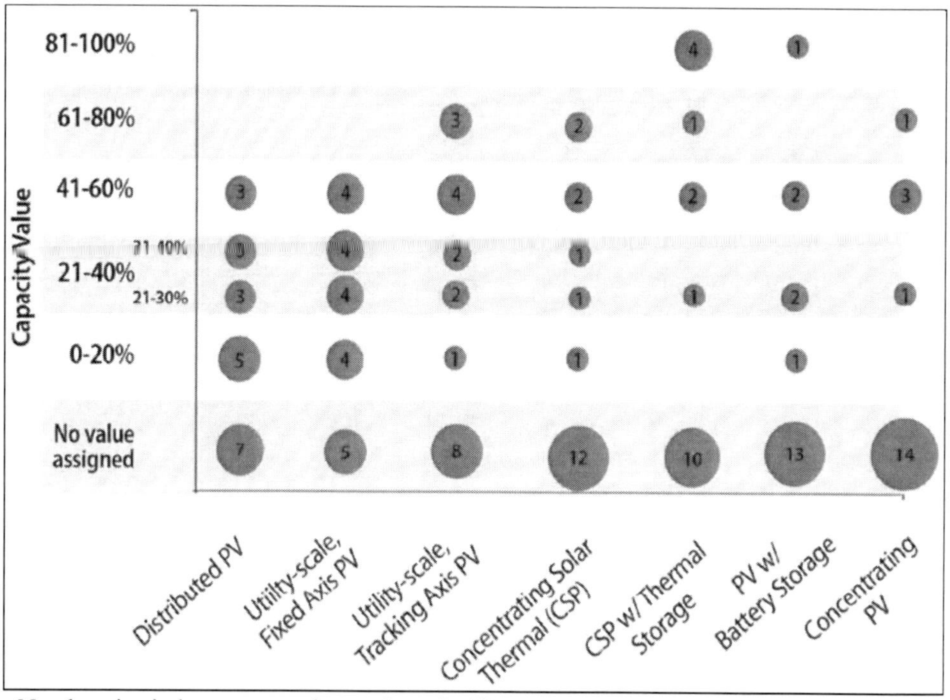

Note: Numbers in circles represent the number of utility responses.

Total number of responses: 21

b) What are these capacity values based on?

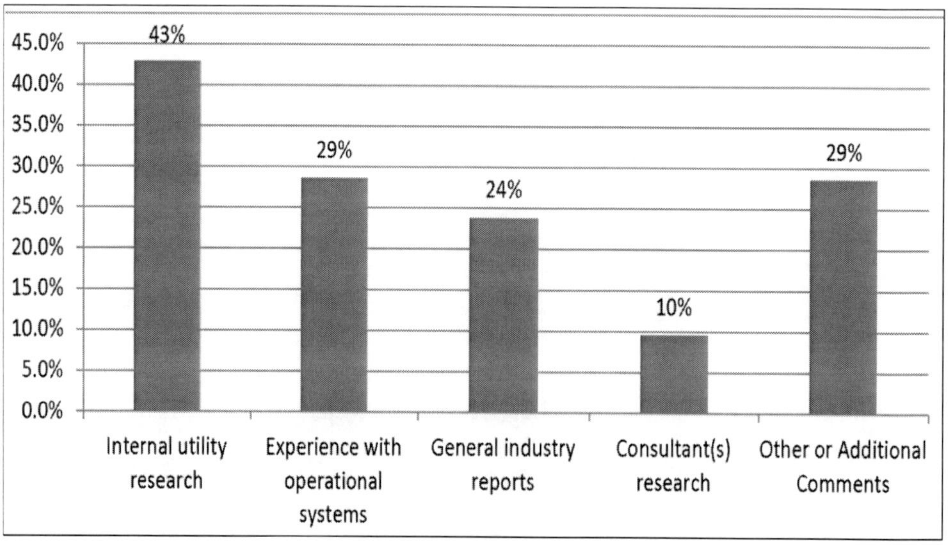

Total number of responses: 21

Other or additional comments:

- Regional Transmission Organization
- We will continue to analyze PV levels of generation, PV ability to meet electricity demand, etc. to determine the capacity value of PV.
- Initially RFP's. Thereafter, we gather actual solar operational production data to track against estimates. If needed, we will adjust our capacity values accordingly.

c) What methodology was utilized to calculate the capacity value?

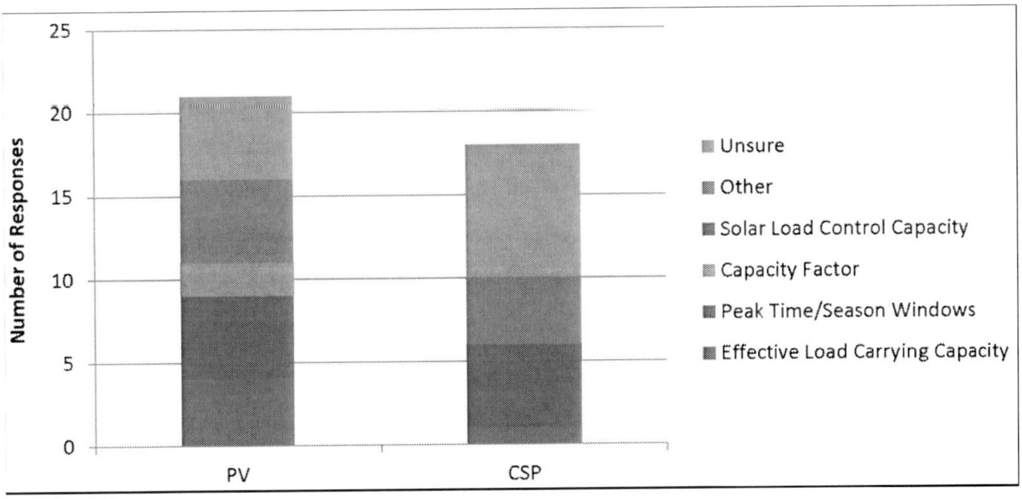

Total number of responses: PV – 21; CSP – 18

9. (a) Do you incorporate an integration cost adder for solar resources?

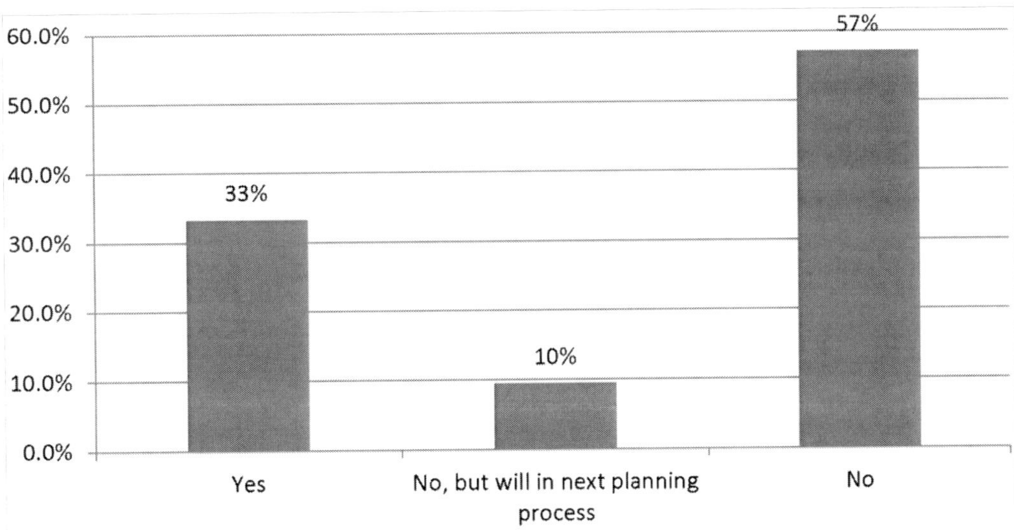

Total number of responses: 21

(b) As solar PV is considered a non-firm resource, does your utility explicitly add a firming cost for valuation purposes?

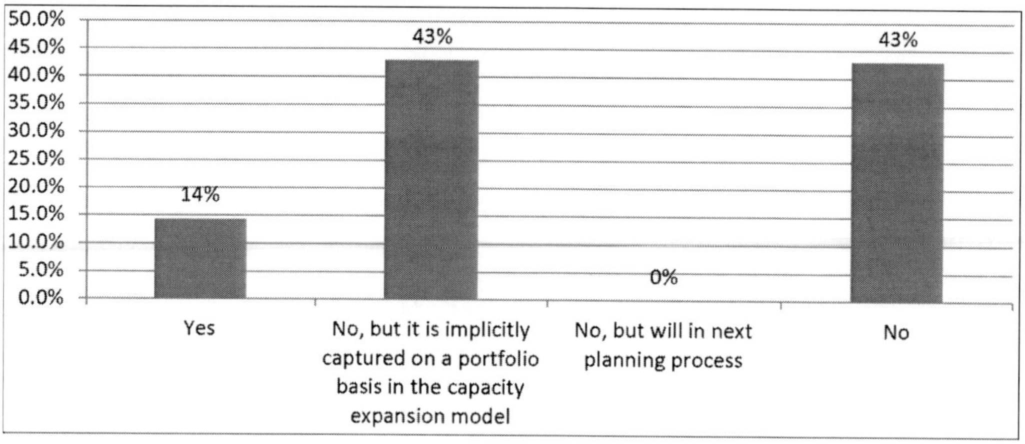

Total number of responses: 21

If yes, please describe the methodology/approach:

- Estimate at this point, as we have little experience.
- Indirectly, by applying coincident capacity value
- Cost of natural gas back up is used
- It varies based on valuation method

REFERENCES

Aliff, Gregory. (2013). "The Math Does Not Lie. Factoring the Future of the U.S. Electric Power Industry." Deloitte Center for Energy Solutions. http://www.deloitte.com/assets/Dcom-UnitedStates/Local%20Assets/Documents/Energy_us_er/us_er_Deloitte MathDoesNotLie_10 2212.pdf.

Ardani, Kristin., Barbose, Galen., Margolis, Robert., Wiser, Ryan., Feldman, David. & Ong, Sean. (2012). *Benchmarking Non-Hardware Balance of System (Soft) Costs for U.S. Photovoltaic Systems Using a Data-Driven Analysis from PV Installer Survey Results.* NREL/TP-7A20-56806; DOE/GO-102012-3834. http://www.nrel.gov/docs/fy13osti/56806.pdf.

Augustine, C., Bain, R., Chapman, J., Denholm, P., Drury, E., Hall, D. G., Lantz, E., Margolis, R., Thresher, R., Sandor, D., Bishop, N. A., Brown, S. R., Cada, G. F. & Felker, F. (2012). "*Renewable Electricity Futures Study.* Volume *2*: *Renewable Electricity Generation and Storage Technologies.* (Volume *2* of 4)." Golden, CO: National Renewable Energy Laboratory. NREL/TP-6A20-52409-2. http://www.nrel.gov/docs/fy12osti/52409-2.pdf.

Barbose, G. (2013). "Renewables Portfolio Standards in the United States: A Status Update." Presentation to the National Conference of State Legislatures. May 2, 2013. http://www.cleanenergystates.org/assets/2013-Files/RPS/Barbose-RPS-Presentation-NCSL- Spring-2013.pdf.

Barbose, Galen L., Naïm, Darghouth., Samantha, Weaver. & Ryan, H. Wiser. (2013). "Tracking the Sun VI: An Historical Summary of the Installed Price of Photovoltaics in the United States from 1998 to 2012." Berkeley: Lawrence Berkeley National Laboratory. http://emp.lbl.gov/sites/all/files/lbnl-6350e.pdf.

California Independent System Operator. (CA ISO 2013). "Renewables: Graph shows aggregated output from renewables connected to the ISO grid." California ISO website: Today's Outlook, accessed August 2013 at http://www.caiso.com/Pages/Todays Outlook.aspx.

Database of State Incentives for Renewables and Efficiency. (DSIRE 2013). "RPS Policies with Solar/DG Provisions." Accessed July 16, at http://www.dsireusa.org/documents/summarymaps/Solar_DG_RPS_map.pdf.

Denholm, P., Jorgenson, J., Hummon, M., Jenkin, T., Palchak, D., Kirby, B., Ma, O. & O'Malley, M. (2013). *Value of Energy Storage for Grid Applications*. NREL/TP-6A20-58465. Golden, CO: National Renewable Energy Laboratory. http://www.nrel.gov/docs/fy13osti/58465.pdf.

Denholm, P. (2012). "*Energy Storage to Reduce Renewable Energy Curtailment*." Prepared for the 2012 IEEE Power and Energy Society General Meeting, 22-26 July 2012. NREL/CP- 6A20-53533. http://dx.doi.org/10.1109/PESGM.2012.6345450.

Denholm, P. (2011). "*Enabling Technologies for High Penetration of Wind and Solar Energy*." Prepared for ASME 2011 5th International Conference on Energy Sustainability & 9th Fuel Cell Science, Engineering and Technology Conference (ESFuelCell2011), 7-10 August 2011. Paper No. ES2011-54500. NREL/CP-6A20-50868. http://dx.doi.org/10.1115/ES2011-54500.

Denholm, P. & Hand, M. (2011). "Grid Flexibility and Storage Required to Achieve Very High Penetration of Variable Renewable Electricity." *Energy Policy*, (*39*:3), March 2011; 1817-1830. NREL/JA-6A20-49400. http://dx.doi.org/10.1016/j.enpol.2011.01.019

Drury, E., Brinkman, G., Denholm, P., Margolis, R. & Mowers, M. (2012). "*Exploring Large- Scale Solar Deployment in DOE's SunShot Vision Study*." Prepared for 38th IEEE Photovoltaic Specialists Conference, 3-8 June 2012. NREL/CP-6A20-55372. http://dx.doi.org/10.1109/PVSC.2012.6317716.

Edison Electric Institute. (EEI 2013). "*Disruptive Challenges: Financial Implications and Strategic Responses to a Changing Retail Electric Business*." Prepared by Peter Kind of the Energy Infrastructure Advocates for the Edison Electric Institute. January 2013. http://www.eei.org/ourissues/finance/Documents/disruptivechallenges.pdf.

Energy Information Administration. (EIA 2013). "*Annual Energy Outlook 2013*." Washington, D.C.: U.S. Department of Energy. DOE/EIA-0383(2013); supporting data to Figure 79, accessed August 2013, http://www.eia.gov/forecasts/aeo/pdf/0383(2013).pdf

Feldman, David., Barbose, Galen., Margolis, Robert., Wiser, Ryan., Darghouth, Naïm. & Goodrich, Alan. (2012). *Photovoltaic (PV) Pricing Trends: Historical, Recent, and Near-Term Projections*. NREL/TP-6A20-56776; DOE/GO-102012-3839. http://www.nrel.gov/docs/fy13osti/56776.pdf.

Hoff, T., Perez, R., Ross, J. P. & Taylor, M. (2008). *Photovoltaic Capacity Valuation Methods*. Solar Electric Power Association Report #02-08. http://www. Solar-electricpower.org/media/84207/sepa%20pv%20capacity.pdf.

Krishanmoorthy, Bart., Mike, Taylor. & Becky, Campbell. (2013). *2012 SEPA Utility Solar Rankings*. Solar Electric Power Association Report #02-13. http://www. solarelectricpower.org/media/324121/final-2012-top-10-report-v2.pdf.

Lehr, Ron. & Binz, Ron. (2013). "*New Utility Business Models: Implications of a High-Penetration Renewable Energy Future*." Prepared for the session "New utility model for a low load growth/high DG future" of the Joint Meeting of Committee on Regional Electric Power Cooperation and State-Provincial Steering Committee, April 8-10, 2013. http://www.westgov.org/wieb/meetings/crepcsprg2013/briefing/Lehr-Binz.pdf.

Lew, D., Brinkman, G., Ibanez, E., Florita, A., Heaney, M., Hodge, B. M., Hummon, M., Stark, G., King, J., Lefton, S. A., Kumar, N., Agan, D., Jordan, G. & Venkataraman, S. (2013). "*Western Wind and Solar Integration Study Phase 2*." NREL/TP-5500-55588. Golden, CO: National Renewable Energy Laboratory. http://www.nrel.gov/docs/ fy13osti/55588.pdf.

Lew, D., Brinkman, G., Kumar, N., Besuner, P., Agan, D. & Lefton, S. (2012). "*Impacts of Wind and Solar on Fossil-Fueled Generators*." Preprint. Prepared for the 2013 IEEE Power and Energy Society General Meeting. NREL/CP-5500-53504. Golden, CO: National Renewable Energy Laboratory. http://www.nrel.gov/docs/fy12osti/53504.pdf.

Lew, Debra., Nicholas, Miller., Kara, Clark., Gary, Jordan. & Zhi, Gao. (2010). *Impact of High Solar Penetration in the Western Interconnection*. NREL/TP-5500-49667. Golden, CO: National Renewable Energy Laboratory. http://www.nrel.gov/docs/fy11osti/ 49667.pdf.

Madaeni, S. H., Sioshansi, R. & Denholm, P. (2013). "Comparing Capacity Value Estimation Techniques for Photovoltaic Solar Power." *IEEE Journal of Photovoltaics*, (*3*:1); January 2013; 407-415. NREL/JA-6A20-54703. http://dx.doi.org/10.1109/JPHOTOV.2012. 2217114.

Mai, Trieu. (et al. – TBD). (2014 Forthcoming – a quantification of the impact of modeling approaches and assumptions on optimal solar deployment levels in utility planning processes). For more information, please contact Trieu Mai (Trieu.Mai@nrel.gov).

Mai, T., Drury, E., Eurek, K., Bodington, N., Lopez, A. & Perry, A. (2013). *Resource Planning Model: An Integrated Resource Planning and Dispatch Tool for Regional Electric Systems*. NREL/TP-6A20- 56723. Golden CO: National Renewable Energy Laboratory. January 2013. http://www.nrel.gov/docs/fy13osti/56723.pdf.

Mai, T., Wiser, R., Sandor, D., Brinkman, G., Heath, G., Denholm, P., Hostick, D. J., Darghouth, N., Schlosser, A. & Strzepek, K. (2012). *Renewable Electricity Futures Study: Volume 1: Exploration of High-Penetration Renewable Electricity Futures*. NREL/TP-6A20- 52409-1. Golden, CO: National Renewable Energy Laboratory. http://www.nrel.gov/analysis/re_futures.

Milligan, M., Ela, E., Hein, J., Schneider, T., Brinkman, G. & Denholm, P. (2012). *Renewable Electricity Futures Study: Volume 4:Bulk Electric Power Systems: Operations and Transmission Planning*." NREL/TP-6A20-52409-4. Golden, CO: National Renewable Energy Laboratory. http://www.nrel.gov/docs/fy12osti/52409-4.pdf.

Mills, A. & Wiser, R. (2012). *An Evaluation of Solar Valuation Methods Used in Utility Planning and Procurement Processes*. LBNL-5933E. Berkeley, CA: Lawrence Berkeley National Laboratory. http://emp.lbl.gov/sites/all/files/lbnl-5933e_0.pdf.

Newcomb, James., Lacy, Virginia. & Hansen, Lena. (2013). "New Business Models for the Distribution Edge: The Transition from Value Chain to Value Constellation" Boulder, CO: Rocky Mountain Institute Electricity Innovation Lab (e-lab). April 2013. www.rmi.org/new_business_models.

Palmintier, B. & M. Webster. (2011). "*Impact of Unit Commitment Constraints on Generation Expansion Planning with Renewables*." Prepared for 2011 IEEE Power and Energy Society General Meeting, Detroit, Institute of Electrical and Electronics Engineers.http://web.mit.edu/b_p/pdf/Palmintier%20and%20Webster%20-%202011% 20-%20Impact%20of%20Unit%20Commitment%20Constraints%20on%20Generation %20Plann ing%20with%20Renewables.pdf.

Perez, R., Margolis, R., Kmiecik, M., Schwab, M. & Perez, M. (2006). "*Update: Effective Load-Carrying Capability of Photovoltaics in the United States*." Preprint. Prepared for the 2006 American Solar Energy Society 35th Annual Conference. NREL/CP-620-40068. Golden, CO: National Renewable Energy Laboratory. June 2006. http://www. nrel. gov/docs/fy06osti/40068.pdf.

PVsyst. (2013). "*General description of the PVsyst Software*," PVsyst 6 Help file, accessed August, http://files.pvsyst.com/help/index.html?general_descr.htm.

PVWatts. (2013) "PVWatts" *Renewable Resource Data Center*, accessed August, http://www.nrel.gov/rredc/pvwatts.

Rábago, Karl., Libby, Leslie., Harvey, Tim., Norris, Benjamin. & Hoff, Thomas. (2012). *Designing Austin Energy's Solar Tariff Using a Distributed PV Value Calculator*. Self-published by Karl Rábago and Clean Power Research. Accessed July 9, 2013 at http://rabagoenergy.com/files/value-of-solar-rate.pdf.

State and Local Energy Efficiency Action Network. (SEE Action 2011). Washington, D.C.: U.S. Department of Energy SEE Action Program. http://www1.eere.energy.gov/ seeaction/pdfs/ratepayer_efficiency_irpportfoliomanagement.pdf.

Solar Electric Power Association (SEPA). (2013). *Centralized Solar Projects Bulletin: Member-Only News and Information*. Q2 2013 edition. August 2013. http://www. solarelectricpower.org/media/351408/sepa-bulletin-q2-2013-final.pdf.

Sioshansi, R., Denholm, P. & Jenkin, T. (2012). "Market and Policy Barriers to Deployment of Energy Storage." *Economics of Energy and Environmental Policy*, (*1*:2); March 2012; pp. 47- 63; NREL/JA-6A20-53820. http://dx.doi.org/10.5547/2160-5890.1.2.4.

Wilson, Rachel. & Bruce, Biewald. (2013). "*Best Practices in Electric Utility Integrated Resource Planning*." Prepared by Synapse Energy Economics for the Regulatory Assistance Project. June 2013. http://www.raponline.org/document/download/id/6608 .

Wilson, Rachel. & Paul, Peterson. (2011). "*A Brief Survey of State Integrated Resource Planning Rules and Requirements*." Prepared by Synapse Energy Economics for the Clean Skies Foundation. April 28. http://www.cleanskies.org/wp- content/uploads/2011/ 05/ACSF_IRP-Survey_Final_2011-04-28.pdf.

Wyoming Public Service Commission. (WY PSC 2013). "Commission Guidelines Regarding Electric IRP." Accessed August 2013 at http://psc.state.wy.us/pscdocs/dwnload/ ElectricIRPGuidelines7-10.pdf.

End Notes

[1] This report considers distributed solar PV, utility-scale solar PV with and without tracking, concentrated solar power with and without storage, PV with battery storage, and concentrating PV. Because distributed PV and fixed- axis utility-scale PV are the dominant applications in the market, these technologies are the main focus.

[2] For the purposes of this report, "utility" refers to any entity that engages in long-term supply- and demand-side planning to serve the load requirements of its customers. This typically applies to vertically-integrated utilities, which may include investor-owned utilities (IOUs), municipal utilities, and electric cooperatives.

[3] Based on interviews and questionnaire results, it appears that most – but not all – utilities take the steps detailed here. For each step, utilities use different levels of detail, models or alternative calculation methodologies. The differences in the baseline supply planning processes were stark enough to capture the different methodologies used, as they provide a basis for including solar in long-term planning.

[4] For simplicity, the terms Monte Carlo simulations and stochastic modeling are used interchangeably in this report.

[5] Utility resource planners interviewed in this report routinely referred to the percentage of nameplate assigned to solar or other variable resources as its "capacity value". The authors chose to use this term rather than "capacity credit", which is also an acceptable industry term for this characteristic.

[6] These utilities either have no or virtually no solar on their systems today, or have specifically stated that they have not determined the capacity value assigned to solar at this time.

[7] Although the resource planning process and resulting documents are not part of ratemaking procedures, utility supply planning does occur within the broader context of utility decision-making and is impacted by the utility's overarching goals. As such, during the discussions of resource planning, the issue of the impacts of distributed generation on utility revenues and the implications for utility planning processes arose. Here we present only the results of our discussions with utility representatives, and do not holistically explore the issue of utility revenue loss resulting from distributed generation. This key question indicates that further research is needed.

[8] For the purposes of this report, the authors use integrated resource plans, integrated resource planning, and long-term planning interchangeably.

[9] This report considers distributed solar PV, utility-scale solar PV with and without tracking, concentrated solar power with and without storage, PV with battery storage, and concentrating PV. Because distributed solar PV and utility-scale PV are the dominant applications in the market, these technologies are the main focus.

[10] The Joint Resource Planner's Forum, sponsored by the Western Governors' Association, has also provided a forum for information exchange for planners in the U.S. West. See: http://westgov.org/newsite/crepc/past-meetings/.

[11] Individual utility responses are not provided in this report.

[12] Some utilities reported similar planning processes across multiple affiliated companies, and some utilities

[13] To illustrate an example of a successful and thorough process, a case study covering Tennessee Valley Authority's recent IRP is included at Appendix A.

[14] Load growth is highly region-specific, as it depends in large part upon a utility's local economic outlook and population growth.

[15] All values listed in MW-ac.

[16] The preceding statement is specific to TMY2 data, which is based on solar data from 1961–1990, and covers 239 sites. TMY3 (1991-2005) is available for 1,020 locations, but is used less frequently due to the smaller sample size.

[17] The SunShot Initiative's target is a 75% reduction in installed costs by the year 2020, down to $1.50/watt for utility-scale systems.

[18] Solar generation with storage, whether from battery or thermal technologies, can be considered dispatchable.

[19] Additional information on capacity valuation approaches can be found in: Hoff et al. 2008

[20] These utilities either have no or virtually no solar on their systems today, or have specifically stated that they have not determined the capacity value assigned to solar at this time.

[21] Same comment as Footnote 20.

[22] Table 6 "Assumed integration costs used by LSEs to adjust production costs for portfolios with solar." One important caveat is that few utilities have done system-specific studies for integration costs. Several rely upon studies done for wind integration and simply use the same $/MWh cost for solar.

[23] All utilities interviewed were interested in PV. CSP was of interest to a few utilities, but is not a viable option for others due to the quality of resource, land constraints, etc.

[24] Correlation with peak demand is also highly penetration-specific. Section 3 of this report discusses the impact of higher penetration of solar PV, and how the incremental capacity value of the next system falls.

[25] Deferred transmission and distribution (T&D) costs are often discussed as another benefit of solar generation; however, during the report's interview process, utilities either did not mention deferred T&D costs as a benefit, or stated that it may be a benefit but it had not been quantified to date.

[26] This solar ramp can be viewed every day on the California ISO website (second chart), http://www.caiso.com/Pages/TodaysOutlook.aspx.

[27] These profiles would also be correlated to the load forecast, which is already typically part of a utility's Monte Carlo simulations.

[28] In 2014, NREL has several activities to analyze this space more thoroughly including quantifying methodologies for valuing costs and benefits of solar, utility operationalization of value of solar in program design and implementation, and examining DG solar issues from the regulatory perspective. Contact Karlynn Cory (Karlynn.cory@nrel.gov) if you have questions or would like to get involved.

[29] Most retail rates are recovered as per kilowatt-hour (kWh) payments (e.g., ¢/kWh). The fewer customers that make the kWh payments, the more the remaining customers have to pay to cover the system's transmission and distribution costs. It can be a self-defeating cycle where rising rates make it more attractive for the customers to consider on-site generation, so more of them sign up to self-generate, which results in more rate increases for the remaining customers. There is also a concern of cross-subsidization where the few who participate in on-site generation programs (generally more affluent homeowners) are subsidized by the remaining customers (generally middle- and low-income ratepayers).

[30] Another option that is covered in the literature, but not in the scope of the utility discussions, is to consider overhauling cost of service ratemaking used today. Some market analysts question the usefulness of cost of service ratemaking in the future, when the electric sector will be more complex with a wider variety of generation technologies, at the behind-the-meter distributed level, the wholesale distributed level (on the utility side of the meter), and at the transmission level. If regulatory approval could be secured, it may be possible to depart from the traditional model of cost of service ratemaking (Aliff 2013, EEI 2013, and Newcomb et al. 2013).

[31] In some states there are rules that are intended to prevent a utility giving itself an unfair advantage by favoring its generation resources on its transmission resources.

[32] For more information on this ongoing work, please contact Trieu Mai (Trieu.Mai@nrel.gov).

[33] For additional information on this decreasing capacity value, please see: Perez et al. 2006.

[34] Adapted from EEI's Glossary of Electric Industry Terms, http://www.eei.org/resourcesandmedia/products/glossry_electerm/Pages/default.aspx

[35] Quoted from: http://www.solarbuzz.com/resources/glossary

[36] Quoted from EEI's Glossary of Electric Industry Terms, http://www.eei.org/resourcesandmedia/products/glossry_electerm/Pages/default.aspx

[37] Adapted from http://www.renewableenergyworld.com/rea/news/article/2009/10/how-to-compare-power-generation-choices

[38] Quoted from NREL's Energy Analysis description of the Regional Energy Deployment System model (ReEDS), http://www.nrel.gov/analysis/reeds/description.html

[39] Quoted from OpenEI: http://en.openei.org/wiki/Definition:Capacity_factor

[40] Adapted from http://www.ise.osu.edu/isefaculty/sioshansi/papers/pv_cv.pdf

[41] Adapted from http://energy.gov/oe/technology-development/energy-storage

[42] Quoted from U.S. Environmental Protection Agency: http://www.epa.gov/climatechange/ghgemissions/gases.html

[43] Adapted from NREL: http://www.nrel.gov/docs/fy12osti/54527.pdf

[44] Adapted from: http://www.eia.gov/tools/faqs/faq.cfm?id=107&t=3

[45] Quoted from http://www1.eere.energy.gov/tribalenergy/guide/glossary.html

[46] Quoted from http://www.risk.net/energy-risk/glossary/2040393/intermediate-generation-cycling-generation

[47] Adapted from http://www.renewableenergyworld.com/rea/news/article/2009/10/how-to-compare-power-generation-choices

[48] Adapted from RMI: http://www.rmi.org/RFGraph-load_duration_curve

[49] Quoted from Investopedia: http://www.investopedia.com/terms/m/montecarlosimulation.asp

[50] Adapted from Wikipedia: http://en.wikipedia.org/wiki/Monte_Carlo_method

[51] Adapted from Adopted from http://www.renewableenergyworld.com/rea/news/article/2009/10/how-to- compare-power-generation-choices

[52] Quoted from: http://www.eei.org/resourcesandmedia/products/glossry_electerm/Pages/default.aspx

[53] Adapted from Pennsylvania code: http://www.pacode.com/secure/data/052/chapter57/subchapktoc.html

[54] Quoted from ERCOT: http://www.ercot.com/content/meetings/lts/keydocs/2011/0110/Production_Cost_Modeling_Presentation_10JAN2011.pdf

[55] Adapted from: http://en.wikipedia.org/wiki/Power_Purchase_Agreement

[56] Quoted from the EPA: http://www.epa.gov/greenpower/gpmarket/rec.htm

[57] Quoted from Weather Analytics: http://www.weatheranalytics.com/weather-products/simulation-and- modeling/tmy-files/

[58] In general, as long as the long-term plan remains in line with the outcomes of the annual planning cycle, TVA does not deem it necessary to update the IRP. Although originally scheduled to begin in 2015, TVA will begin updating the IRP in late 2013, with a completion date of 2015. This effort is in direct response to stakeholder requests to review the company's plans in light of changing natural gas prices and other key uncertainties. The company has already begun to reach out to stakeholders, including the solar industry, to continue to improve the methods through which solar can best be considered within the IRP process.

[59] Determining which outcomes were the most desirable (i.e. what were the indicators of a successful portfolio) was challenging but critical.

[60] See TVA 2011 IRP, Chapter 6, pages 103-1 for details on the scorecard evaluation.

In: Incorporating Solar Technologies in the Utility Resource ... ISBN: 978-1-63117-009-6
Editor: Sandra M. Grosso

Chapter 2

AN EVALUATION OF SOLAR VALUATION METHODS USED IN UTILITY PLANNING AND PROCUREMENT PROCESSES*

Andrew Mills and Ryan Wiser

ABSTRACT

As renewable technologies mature, recognizing and evaluating their economic value will become increasingly important for justifying their expanded use. This report reviews a recent sample of U.S. load-serving entity (LSE) planning studies and procurement processes to identify how current practices reflect the drivers of solar's economic value. In particular, we analyze the LSEs' treatment of the capacity value, energy value, and integration costs of solar energy; the LSEs' treatment of other factors including the risk reduction value of solar, impacts to the transmission and distribution system, and options that might mitigate solar variability and uncertainty; the methods LSEs use to design candidate portfolios of resources for evaluation within the studies; and the approaches LSEs use to evaluate the economic attractiveness of bids during procurement.

We found that many LSEs have a framework to capture and evaluate solar's value, but approaches varied widely: only a few studies appeared to complement the framework with detailed analysis of key factors such as capacity credits, integration costs, and tradeoffs between distributed and utility-scale photovoltaics. Full evaluation of the costs and benefits of solar requires that a variety of solar options are included in a diverse set of candidate portfolios. The design of candidate portfolios evaluated in the studies, particularly regarding the methods used to rank potential resource options, can be improved. We found that studies account for the capacity value of solar, though capacity credit estimates with increasing penetration can be improved. Furthermore, while most LSEs have the right approach and tools to evaluate the energy value of solar, improvements remain possible, particularly in estimating solar integration costs used to adjust energy value. Transmission and distribution benefits, or costs, related to solar are rarely included in studies. Similarly, few LSE planning studies can reflect the full range of potential benefits from adding thermal storage and/or natural gas augmentation to

* This is an edited, reformatted and augmented version of the Office of Energy Efficiency & Renewable Energy, DEAC02-05CH1123, dated December 2012.

concentrating solar power plants. Finally, the level of detail provided in requests for proposals used in procurement is not always sufficient for bidders to identify the most valuable technology or configurations to the LSE.

ACRONYMS AND ABBREVIATIONS

APS	Arizona Public Service
AS	Ancillary service
CA IOU	California Investor-Owned Utility
CAISO	California Independent System Operator
CCGT	Combined cycle gas turbine
CPUC	California Public Utilities Commission
CSP	Concentrating solar power
CT	Combustion turbine
DG	Distributed generation
ELCC	Effective load-carrying capability
IID	Imperial Irrigation District
IRP	Integrated resource plan
LADWP	Los Angeles Department of Water and Power
LCBF	Least-cost, best-fit
LOLP	Loss of load probability
LSE	Load-serving entity
LTPP	Long-Term Procurement Plan
NGST	Natural gas steam turbine
NPCC	Northwest Power and Conservation Council
NQC	Net qualifying capacity
NREL	National Renewable Energy Laboratory
O&M	Operations and maintenance
PG&E	Pacific Gas and Electric
PGE	Portland General Electric
PNM	Public Service of New Mexico
PPA	Power purchase agreement
PSCo	Public Service of Colorado
PV	Photovoltaics
PVRR	Present value of the revenue requirement
RA	Resource adequacy
REC	Renewable energy certificate
RETI	Renewable Energy Transmission Initiative
RFP	Request for proposals
RPS	Renewables portfolio standard
SCPPA	Southern California Public Power Authority
SEPA	Solar Electric Power Association
T&D	Transmission and distribution
TEP	Tucson Electric Power
TWh	Terawatt-hour

WECC	Western Electricity Coordinating Council
WREZ	Western Renewable Energy Zone Initiative

EXECUTIVE SUMMARY

Introduction

Recent declines in the cost of photovoltaic (PV) energy, increasing experience with the deployment of concentrating solar power (CSP), the availability of tax-based incentives for solar, and state renewables portfolio standards (RPS) (some with solar-specific requirements) have led to increased interest in solar power among U.S. load-serving entities (LSEs). This interest is reflected within LSE planning and procurement processes and in a growing body of literature on the economic value of solar energy within utility portfolios. This report identifies how current LSE planning and procurement practices reflect the drivers of solar's economic value identified in the broader literature. This comparison can help LSEs, regulators, and policy makers identify ways to improve LSE planning and procurement.

The report reviews 16 planning studies and nine documents describing procurement processes created during 2008–2012 by LSEs interested in solar power (Table ES1). We first summarize the typical approach used by LSEs in planning studies and procurement processes. We then analyze the LSEs' treatment of the capacity value, energy value, and integration costs of solar energy; the LSEs' treatment of other factors including the risk reduction value of solar, impacts to the transmission and distribution system, and options that might mitigate solar variability and uncertainty; the methods LSEs use to design candidate portfolios of resources for evaluation within the studies; and the approaches LSEs use to evaluate the economic attractiveness of bids during procurement. We offer several recommendations that could help LSEs improve planning studies and procurement processes.

Summary of Steps Used by LSEs in Planning Studies and Procurement Processes

Many of the LSEs followed a similar set of steps that began with an assessment of demand forecasts, generation options, fuel price forecasts, and regulatory requirements over a planning horizon. Based on this assessment, LSEs created candidate resource portfolios that satisfy these needs and regulatory requirements. These candidate portfolios were typically created using one of three methods:

- Manual creation based on engineering judgment or stakeholder requests
- Creation using capacity-expansion models based on deterministic future assumptions
- Creation using an intermediate approach in which resource options are ranked according to metrics defined by each LSE

The present value of the revenue requirement (PVRR) of candidate portfolios was then evaluated in detail. The PVRR of each portfolio was based primarily on the capital cost of each portfolio and the variable cost of dispatching each portfolio to maintain a balance between supply and demand over the planning period. The variable cost was commonly evaluated by simulating the dispatch of the portfolio using a production cost model. Many LSEs used scenario analysis or Monte-Carlo analysis (or some combination of both) to evaluate the exposure of each portfolio to changes in uncertain factors such as fossil-fuel prices, demand, or carbon dioxide prices. LSEs then chose a preferred portfolio based on the relative performance of the candidate portfolios. The preferred portfolio was often determined by balancing a desire for both low costs and low risks. During procurement, LSEs often solicited bids for resources that matched the characteristics of resources identified in the preferred portfolio.

Solar Technologies Considered in Planning and Procurement

Among our sample, many LSEs considered PV and CSP with or without thermal storage or natural gas augmentation.[1] The PV technologies considered by LSEs were not always described in detail. When they were described, LSEs typically considered fixed PV or single-axis tracking PV; some also distinguished between distributed and utility-scale PV. One LSE considered a PV plant coupled with a lead-acid battery. The CSP technology was usually based on a parabolic trough or a solar power tower configuration. One LSE considered a solar chimney, and another LSE considered a solar thermal gas hybrid (a natural gas power plant with solar concentrators that preheat water used in the plant's steam cycle).

Recognition of Solar Capacity Value in Planning Studies

In regions where solar generation is well correlated with periods of high demand, one of the main contributors to solar's economic value is the capacity value. The capacity value of solar reflects the avoided costs from reducing the need to build other capacity resources, often combustion turbines (CTs), to meet peak demand reliably. LSEs usually added sufficient capacity to meet the peak load plus a planning reserve margin in each candidate portfolio. Portfolios that included solar need not include as much capacity from other resources, so solar offset some of the capital cost that would otherwise be included in the portfolio's PVRR. Thus, solar's capacity value was based in part on the capital cost of the avoided capacity resources and the timing of the need for new capacity.

The capacity value of solar was affected by the study methodology. In at least one case, the LSE assumed that the generating resources used for capacity were very "lumpy" (i.e., only available in blocks of 290 MW or greater). As a result, adding a small amount of solar to a portfolio could not change the timing or amount of other capacity resources required; thus, the same amount of CT capacity was needed with or without the inclusion of solar, even though the LSE recognized that some of the solar nameplate capacity could contribute to meeting peak loads. Including capacity resources that are available in smaller size increments—e.g., 50-MW CTs, which were modeled by other LSEs—or modeling the value of selling excess capacity to neighboring LSEs better recognizes solar's capacity value.

Table ES1. Planning studies and procurement practices reviewed in this analysis

Load-serving entity or study author	Planning study (year)	Procurement practices (year)
Arizona Public Service	2012	2011
California IOU Process	2010	2011
Duke Energy Carolinas	2011	-
El Paso Electric	2012	2011
Idaho Power	2011	-
Imperial Irrigation District	2010	-
Los Angeles Department of Water and Power	2011	2012
Northwest Power and Conservation Council	2010	-
NV Energy	2012	2010
PacifiCorp	2011	2010
Portland General Electric	2009	2012
Public Service of Colorado	2011	2011
Public Service of New Mexico	2011	2011
Salt River Project	2010	-
Tri-State Generation and Transmission	2010	-
Tucson Electric Power	2012	-

Estimates of Solar Capacity Credit in Planning Studies and Broader Literature

The primary driver of solar's capacity value is the capacity credit: the percentage of the solar nameplate capacity that can be counted toward meeting the peak load and planning reserve margin. The capacity credit assigned to solar technologies by the LSE determines how much capacity from an alternative resource can be avoided by including solar in a portfolio. For example, a capacity credit of 50% for PV indicates that a 100-MW PV plant can contribute roughly the same toward meeting peak load and the planning reserve margin as a 50-MW CT. Analysis in the literature shows that the capacity credit of solar largely depends on the correlation of solar production with LSE demand, meaning the capacity credit varies by solar technology (e.g., PV vs. CSP with thermal storage), configuration (e.g., single-axis tracking PV vs. fixed PV), and LSE (e.g., summer afternoon peaking vs. winter night peaking). As expected, the capacity credit assigned by LSEs to solar in planning studies varied by technology, configuration, and LSE (Figure ES1). However, few studies appeared to use detailed loss of load probability (LOLP) studies to determine the capacity credit of solar. Instead, most LSEs relied on analysis of the solar production during peak-load periods or assumptions based on rules of thumb. The reliance on assumptions or simple approximation methods to assign a capacity credit to solar may also contribute to much of the variation in capacity credit across studies.

Only one LSE, Arizona Public Service, appeared to account for changes in the capacity credit of solar with increasing penetration. Analysis in the broader literature finds that solar capacity credit decreases with increasing solar penetration, particularly for PV and CSP without thermal storage or natural gas augmentation (Figure ES2). One of the main factors in the literature that distinguishes the economic value of CSP with thermal storage from the economic value of PV and CSP without thermal storage or natural gas augmentation is the ability of CSP with thermal storage to maintain a high capacity credit with increasing penetration. If LSE planning studies do not reflect this difference in capacity credit with increasing penetration, then the difference in economic value among different solar technologies will not be reflected in their planning studies.

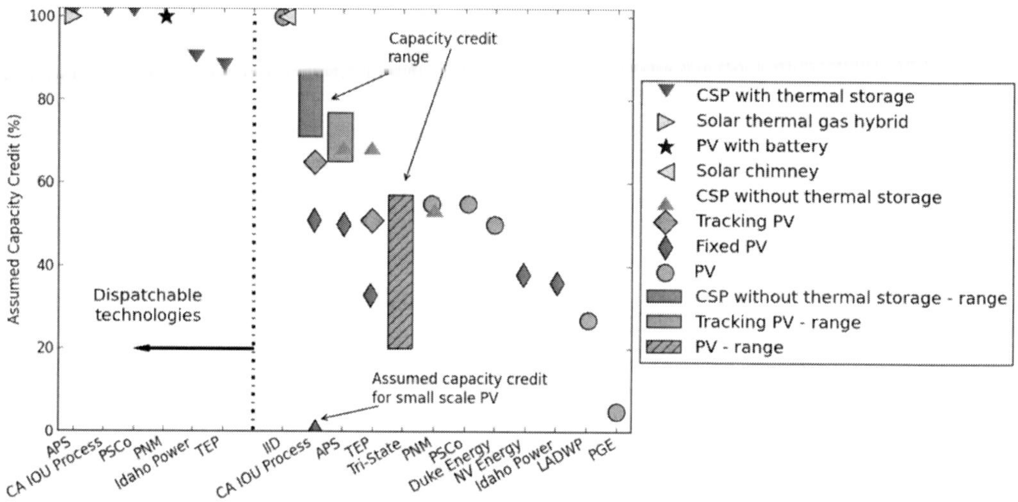

Note: Imperial Irrigation District (IID) appears to assume a 100% capacity credit for PV and a solar chimney. Capacity credit for APS represent capacity credit applied at low penetration level; capacity credit is reduced with higher PV penetration. Range of capacity credits for APS and CA IOU process are based on different plant locations.

Figure ES1. Capacity credits applied by LSEs in planning studies.

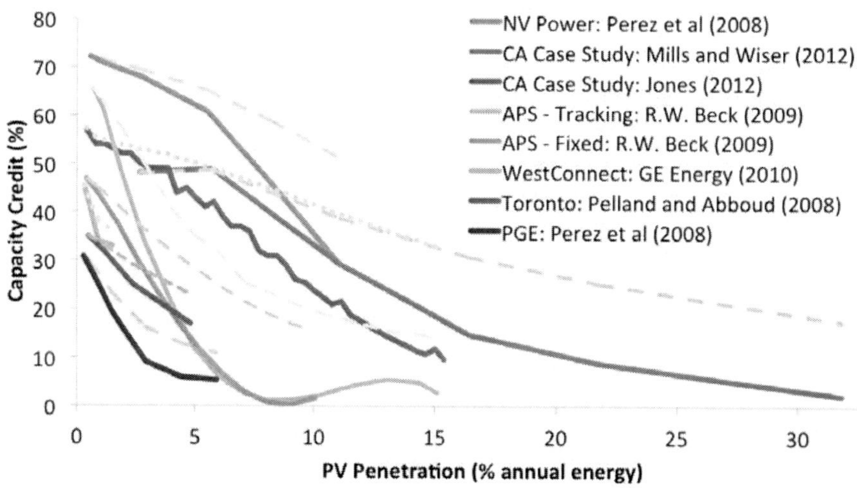

See main text for additional notes.

Figure ES2. PV capacity credit estimates with increasing penetration levels (dashed line is average capacity credit, solid line is incremental capacity credit).

Given the importance of solar's capacity credit for determining economic value and ensuring reliability, LSEs should consider conducting detailed estimates of solar capacity credit. LSEs considering portfolios with large amounts of solar may also need to account for expected changes in the solar capacity credit with increasing penetration.

Evaluation of the Energy Value of Solar Using Production Cost Models

In addition to capacity value, another primary driver of solar's economic value is the energy value. The energy value reflects the reduction in the PVRR from avoiding variable fuel and operational costs from conventional power plants in portfolios with solar. When LSEs evaluate candidate portfolios, they often use production cost models that account for the temporal variation in solar generation, demand, and other resource profiles. Many of the production cost models used by LSEs in planning studies have hourly temporal resolution (either over a one- week period each month or over the full year), and some production cost models account for the various operational constraints of conventional generation. These models appear to account for any benefit from solar generation being correlated with times when plants with high variable costs would otherwise be needed.

The LSEs in our sample that included CSP with thermal storage in candidate portfolios did not describe the approach they used to account for the dispatchability of CSP with thermal storage in the production cost models. In previous analyses, CSP with thermal storage was assumed to operate with a fixed generation profile in which the thermal storage generates as much power as possible in specific, static periods. While this simplified approach may capture some of the benefits of thermal storage, the full benefits to a particular LSE can be better captured by modeling the dispatchability of CSP directly in the production cost model. Compared to thermal storage, natural gas augmentation is relatively easier to model in a production cost model. One LSE described its approach to incorporating natural gas augmentation into its model.

The production cost models used by most LSEs also can account for changes in the energy value as the penetration of solar increases. One key factor in this regard is how LSEs consider the broader wholesale market and the assumptions they make about solar penetration in neighboring markets. If the LSE assumes other regions do not add solar, then selling power to the broader market during times of high insolation and low load may mitigate reductions in the energy value as the penetration of solar increases in the candidate portfolio. Such opportunities may not be available to the same degree, however, if many LSEs in a region simultaneously add solar. LSEs can improve their planning studies by better describing the assumptions and approaches used to account for broader wholesale markets when using production cost models to evaluate candidate portfolios.

Adjusting the Energy Value to Account for Integration Costs

Many LSEs adjust production cost model assumptions or results to account for solar integration costs. Adjustments make sense when there are factors that cannot be represented in the production cost model owing to data or computational limitations. In that case, the adjustments could be tailored to account for the shortcomings of a specific LSE's modeling approach or production cost model. Two studies accounted for solar integration costs by increasing the operating reserve requirement in the hourly production cost model to account for sub-hourly variability and uncertainty that otherwise would be ignored. The increase in operating reserves was based on a separate detailed analysis of sub-hourly variability and uncertainty of solar, wind, and load. Alternatively, other LSEs directly added an estimated integration cost to the production cost model results depending on the amount of solar included in the candidate portfolio. The integration costs for solar added to the production cost model results ranged from $2.5/MWh to $10/MWh. Of the LSEs that used this approach, only one conducted a detailed study of solar integration costs (based on day ahead forecast errors). The remaining LSEs relied on assumptions, results from studies in other regions, or integration cost estimates for wind. Based on the scarcity of detailed analysis of solar integration costs and the wide range of integration cost estimates used in the planning studies, more LSEs should consider carefully analyzing solar integration costs for their system (estimating what is not already captured by their modeling approach) to better justify their assumptions.

Additional Factors Included or Excluded from Planning Studies

Aside from the capacity and energy values, other attributes of solar are often also included in planning studies. The potential risk-reduction benefit of solar, for example, can be accounted for in studies that evaluate the performance of candidate portfolios with and without solar under different assumptions about the future. Transmission and distribution benefits, or costs, related to solar are not often accounted for in LSE studies. In one clear exception, avoided distribution costs were directly accounted for by one LSE in portfolios with distributed PV. In a few other cases, candidate portfolios with solar required less transmission than candidate portfolios with other generation options. The difference in avoided costs between utility-scale solar and distributed PV are not well known, but as more studies provide insight into these differences, LSEs should consider incorporating that information into their planning studies.

A number of LSE planning studies included options that may increase the economic value of solar. Some LSEs included thermal storage or natural gas augmentation with CSP plants, one study considered PV coupled with a lead-acid battery, and another added grid-scale batteries to candidate portfolios with wind and solar (in both cases the additional capital cost of the batteries was too high to reduce the overall PVRR relative to the cases without batteries). Other studies considered a wide range of grid-level storage options without explicitly tying these storage resources to the candidate portfolios with wind or solar. None of the studies appeared to directly consider the role of demand response in increasing the value of solar or directly identify synergies in the capacity credit or integration costs for combinations of wind and solar. Any such synergy in energy value, on the other hand, may have been indirectly accounted for in production cost modeling of candidate portfolios with combinations of wind and solar.

Designing Candidate Portfolios to Use in Planning Studies

While the overall framework used by many of the LSEs for evaluating candidate portfolios appears to capture many (but not all) solar benefits, one important area for improvement is creating candidate portfolios in the first place. The complex interactions between various resource options and existing generation make it difficult to identify which resource options will be most economically attractive. To manage this complexity, a number of LSEs relied on capacity-expansion models to design candidate portfolios, most of which were based on deterministic assumptions about future costs and needs. The LSEs that did not use capacity-expansion models either manually created candidate portfolios based on engineering judgment or stakeholder input or created candidate portfolios by ranking resource options using simplified criteria.

A logical way to rank resources is to estimate the change in the PVRR of a portfolio from including a particular resource in the portfolio and displacing other resources. This change in PVRR is called the "net cost" of a resource since it represents the difference between the cost of adding the resource and the avoided cost from displacing other resources that are no longer needed to ensure the portfolio can meet reliability and regulatory constraints. Since the goal of many planning studies is to minimize the expected PVRR, the resources with the lowest net cost should be added to the portfolio. LSEs in California used a similar approach to identify renewable resource options that were included in their candidate portfolios.

In contrast, a number of LSEs used the levelized cost of energy of resource options along with various adjustments (often based on capacity and integration cost adjustments) to rank resource options. The adjustments, particularly the capacity adjustments, were often not clearly justified and did not always link back to the broader objective of minimizing the expected PVRR. Based on these findings, we recommend that, where possible, LSEs use capacity-expansion models to build candidate portfolios. Improvements in capacity expansion models to account for factors like risk, uncertainty, dispatchability of CSP plants with thermal storage, and operational constraints for conventional generation may be appropriate for some LSEs. If using a capacity- expansion model to build candidate portfolios is not possible, then an approach like the net cost ranking should be considered instead.

Economic Evaluation of Bids in Procurement Processes

Finally, we found that LSE procurement often evaluated the economic attractiveness of bids based on the estimated net cost, but often it was unclear exactly how this net cost was evaluated. The lack of clarity in many procurement documents makes it difficult for a bidder to estimate how various choices it makes in terms of solar technology or configuration will impact the net cost of its bid. The bidder will know how these choices affect the cost side of the bid but often must guess or try to replicate the LSE's planning process to determine how different choices will affect the LSE's avoided costs. LSEs likely could elicit more economically attractive bids by providing as much detail as possible on how the net cost of each bid will be evaluated and the differences in the LSE's avoided costs for different technologies and configurations.

Although this review focused on the valuation of solar in planning and procurement, many of the LSEs are considering other renewable technologies, particularly wind. The lessons learned from this analysis and many of the recommendations apply to the evaluation of other renewable energy options beyond solar.

1. INTRODUCTION

With increased worldwide deployment of solar energy technologies, the cost of generating power from photovoltaics (PV) and concentrating solar power (CSP) has decreased and is expected to decrease further (Barbose et al. 2012, Arvizu et al. 2011, EASAC 2011, Chu and Majumdar 2012). As the cost of solar generation falls, load-serving entities (LSEs), regulators, and policy makers increasingly consider solar generation as one of the many viable options for supplying electricity. For an LSE, solar power provides energy and can satisfy some peak electricity demand in place of conventional generation resources (Hoff 1988, Perez et al. 2008, Pelland and Abboud 2008, Madaeni et al. 2012a, Olson and Jones 2012, Mills and Wiser 2012), helps meet state renewables portfolio standard (RPS) targets (Wiser and Barbose 2008; Wiser et al. 2011), reduces exposure to uncertain fossil fuel and carbon dioxide prices (Bokenkamp et al. 2005, Bolinger and Wiser 2009, Denholm et al. 2009, Perez et al. 2011), and provides ancillary services in the case of CSP with thermal storage and/or natural gas augmentation (EASAC 2011, Madaeni et al. 2012b).

Generally, the primary resource-procurement considerations of an LSE include its needs for capacity and energy, regulatory requirements (state and federal), the relative impact of resources on the LSE's revenue requirement, and the impact of the resource options on the LSE's exposure to future cost uncertainty. Regarding the impact on the revenue requirement, the cost of contracting or building a new plant is in part offset by the costs that the new resource allows the LSE to avoid. These avoided costs are sometimes called the "economic value" of a new generation resource.

Detailed estimates of the economic value of solar in previous analyses show that solar can have high economic value at low penetration levels where there is high coincidence of solar generation and periods of high demand (Borenstein 2008, Lamont 2008, R.W. Beck 2009, Sioshansi and Denholm 2010). At least at low penetration, the economic impact of imperfect forecastability and the need for increased ancillary services with solar appear to be secondary to the higher capacity and energy value (EnerNex 2009, Mills and Wiser 2010, Navigant Consulting et al. 2011, Mills and Wiser 2012). In addition, various studies highlight synergistic effects between combinations of renewable generating technologies such as solar and wind (Denholm and Hand 2011, Nagl et al. 2011, Fripp 2012), PV and CSP with thermal storage (Denholm and Mehos 2011), and PV with storage or demand response (Denholm and Margolis 2007).

Detailed studies also show that the economic value of PV and CSP without thermal storage (or natural gas augmentation) decreases with increasing penetration (Olson and Jones 2012, Mills and Wiser 2012). Increased solar penetration reduces the net load during the day, so eventually the period of peak load net of solar generation shifts into the early evening, even where there is high coincidence of load and insolation. Further, on days with relatively lower load, solar will start to displace generation resources with lower variable costs, such as coal (Denholm et al. 2009, Olson and Jones 2012, Mills and Wiser 2012). At low penetration, the addition of 2–4 hours of thermal storage to CSP appears more valuable than CSP without thermal storage (Madaeni et al. 2012b),[2] although in a separate analysis, CSP plants with 6 hours of thermal storage do not appear to have a significantly greater value (in $/MWh terms) than other solar technologies (Mills and Wiser 2012). On the other hand, the benefit of thermal storage is clear at high penetrations (above about 10% penetration on an annual energy basis) because it helps avoid the otherwise significant decline in value of solar with increasing solar penetration (Mills and Wiser 2012).

This report investigates whether the understanding of solar's economic value from the research literature is reflected in the treatment of solar options in LSE planning studies and procurement processes.[3] We compare methods used by different LSEs, primarily located in the Western and Southern United States. Where possible, we also highlight potential improvements that LSEs (or regulators that oversee them) could make to current planning and procurement methodologies to better reflect the economic value of solar.[4] In particular, we focus on the methods for representing the capacity value,[5] energy value, costs associated with day-ahead forecast errors and ancillary services, and various other factors including transmission and distribution system impacts, measures that might mitigate solar variability and uncertainty and portfolio risk reduction. Although these attributes are not necessarily uniquely identified for solar in LSE planning studies, organizing the discussion around these attributes helps illuminate the many drivers that ultimately affect the attractiveness of resource portfolios that include solar. We also look for indications that LSEs are considering attributes that might increase the value of solar, such as PV tracking, CSP with thermal storage and natural gas augmentation, or synergistic interactions between multiple technologies like PV and CSP or solar and wind. In addition, we review LSE procurement processes to determine how publicly available documents communicate what configurations and technologies are most economically attractive to LSEs and how the processes allow solar developers and equipment manufacturers to communicate to the LSE their technologies' capabilities (e.g., PV tracking, natural gas firing in CSP plant boilers, and integrated storage). This review evaluates the impression that LSE planning and procurement practices focus too much on levelized cost comparisons that ignore considerations of the economic value of solar (e.g., Joskow 2011[6]). Although this report focuses on solar, a sizeable number of its insights apply to the evaluation of other renewable energy options, such as wind.

This report builds on previous analysis of the treatment of renewable energy (Wiser and Bolinger 2006) and carbon regulatory risk (Barbose et al. 2008) in utility resource plans in the western United States, and a survey of the treatment of solar in utility procurement processes (SEPA 2009, Text Box 1). Research into incorporating renewables, other non-conventional technologies, and uncertainty into utility planning has a long history and remains active. Hirst and Goldman (1991), for example, review best practices for integrated resource planning and distinguish it from traditional utility planning. Doherty et al. (2006) examine the impact of wind in generation portfolios, with particular attention to the benefit of increasing the diversity of generation resources, using mean-value portfolio theory, one of the approaches to understanding optimal generation portfolios reviewed in Bazilian and Roques (2008). Jin et al. (2011) and Vithayasrichareon and MacGill (2012) propose refined methods to solve large-scale generation- expansion problems while accounting for uncertainty. For improving capacity-expansion modeling, Shortt et al. (2012) compare the production costs estimated using a simple dispatch- only model with those using a detailed unit-commitment and economic dispatch model with increasing wind penetration in Texas, Ireland, and Finland. They find that the difference in production costs increases with wind penetration due to conventional generation flexibility issues that are only captured in the unit-commitment model.

Note that this study focuses on the planning and procurement ***methodologies*** used by LSEs. Except for assumptions unique to solar, we do not focus on the particular ***assumptions*** made by LSEs, such as fossil fuel price forecasts, carbon cost estimates, or capital cost estimates for conventional plants. However, such assumptions do impact the value of solar and are likely to have just as much impact on LSE planning decisions as the methodological approach. This review also does not focus on the methods used by LSEs to estimate the costs of building or procuring new solar plants. Other studies illuminate current cost trends, drivers, and uncertainties (e.g., IEA 2011, Barbose et al. 2012, EASAC 2011, Goodrich et al. 2012). Although we do not discuss the solar cost assumptions used by LSEs, all LSEs account for the capital costs of the solar technologies considered in their planning studies and procurement processes.

TEXT BOX 1. SEPA SURVEY OF LSE PERCEPTIONS OF THE VALUE OF SOLAR IN PROCUREMENT

Instead of reviewing publicly available planning and procurement documents, as we do in this report, the Solar Electric Power Association (SEPA 2009) directly surveyed LSEs. In response to survey questions about their assessment of solar attributes in procurement, LSEs attached the following qualitative values to solar:

Highest value:

- No emissions of carbon or pollutants
- Carbon offset value

Moderate value:

- Correlation between solar generation and peak hours of utility
- Dispatchability (CSP with storage)
- Elimination of fuel price uncertainty
- Fuel diversification

Lowest value:

- Potential for location close to load
- Minimal water use
- Delay of transmission or distribution investment
- Power factor correction and local voltage support

Section 2 of this report describes the LSE planning and procurement documents we reviewed. Section 3 summarizes the steps LSEs use in planning studies and procurement processes. Section 4 analyzes solar's role in planning and procurement, including discussions of solar capacity values and credits, energy value, integration costs, and other factors as well as the design of candidate portfolios for planning studies and the economic evaluation of bids in procurement processes. Section 5 offers conclusions and recommendations. Following the main text, we provide additional details on the planning and procurement documents we reviewed, LSE planning and procurement practices, and a derivation of net cost based on the objective of minimizing the expected present value of the revenue requirement for a portfolio of resources.

2. LSE Planning and Procurement Documents Reviewed in This Report

In order to understand the current practices used by LSEs to estimate the economic value of solar, we reviewed a sample of 16 integrated resource plans (IRPs) or similar planning documents and nine documents outlining evaluation procedures for procurement (Table 1). These documents were created between 2008 and 2012 and are primarily from LSEs in the western United States that are considering solar power, among other options. Details of the planning and procurement approaches used by each LSE is further documented in Appendix A.

Table 1. Planning studies and procurement practices reviewed in this analysis

Load serving entity or study author	Planning study (year)	Procurement practices (year)
Arizona Public Service	2012	2011
California IOU Process	2010	2011
Duke Energy Carolinas	2011	-
El Paso Electric	2012	2011
Idaho Power	2011	-
Imperial Irrigation District	2010	-
Los Angeles Department of Water and Power	2011	2012
Northwest Power and Conservation Council	2010	-
NV Energy[7]	2012	2010
PacifiCorp	2011	2010
Portland General Electric	2009	2012
Public Service of Colorado	2011	2011
Public Service of New Mexico	2011	2011
Salt River Project	2010	-
Tri-State Generation and Transmission	2010	-
Tucson Electric Power	2012	-

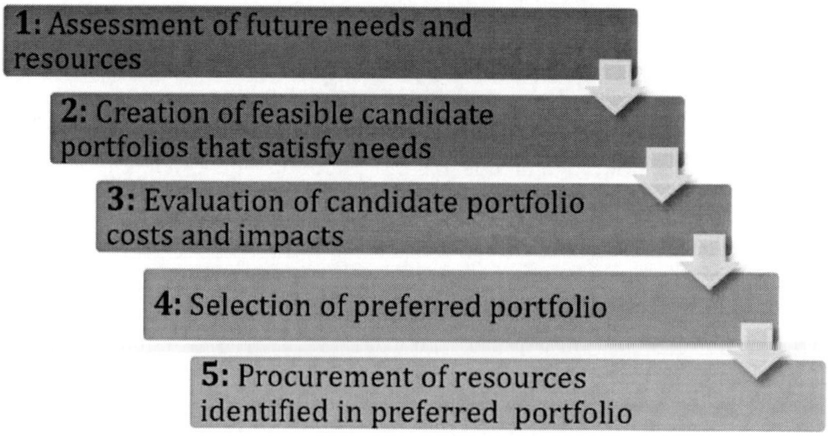

Figure 1. General steps followed by LSEs in planning and procurement.

3. Summary of Steps Used by LSEs in Planning Studies and Procurement Processes

The general planning process adopted by many LSEs followed a similar pattern: the LSE conducted the planning study, identified a preferred plan from the study, and then issued a request for proposals (RFP) for project developers to provide power from resources identified in the preferred portfolio.[8] In this section we briefly describe the structure of the approach used by LSEs in these planning studies and procurement practices. Figure 1 simplifies the varied plans reviewed for this report. Not all LSEs exactly followed these steps: depending on the plan, some steps were not included, multiple steps were bundled into one step, or the order of steps did not follow this same pattern. For details, see the list of planning and procurement documents reviewed at the end of this report.

The most important steps for considering solar are the creation of feasible candidate portfolios (Step 2) and the evaluation of candidate portfolio costs and impacts (Step 3), so our later review emphasizes these steps.[9] In the meantime, the following subsections describe each step and then provide a simple illustration of how the costs and benefits of solar can be evaluated in planning studies.

3.1. Step 1: Assessment of Future Needs and Resources

In the initial stage of their planning studies, the LSEs evaluated expectations and uncertainties for elements that may impact their operations and options in the future. They considered demand forecasts, cost and availability of demand-side management measures, existing generation and contracts, expectations for generator retirement, regulatory and policy constraints, and new generation options, characteristics, and costs. Often a key result of this step was the identification of the gap between existing and planned generating resources in each year and the forecast of peak demand plus a planning reserve margin (Figure 2). This gap describes a constraint that the combination of resources in each candidate portfolio must be able to meet, but it does not describe a decision in terms of what resources can be part of candidate portfolios that satisfy this constraint. In other words, not all resources must contribute equally to meeting this constraint as long as the combination of resources meets the constraint.

Expected future demand levels impacted the degree to which an LSE may need additional capacity and what time of day and year demand was expected to be highest. Understanding future demand included forecasting load growth, understanding trends in energy efficiency, identifying options available for demand-side management (both in terms of LSE-led energy efficiency programs and demand-response programs), and adhering to regulatory requirements for implementing demand-side measures. The level of detail and sophistication used to assess demand-side resources in the planning studies varied across the LSEs.

The supply side included both existing and new resources. These supply-side resource options determined in part the costs that could be avoided by including solar in a portfolio instead of other supply options. Costs for existing generation included assessments of future operations and maintenance (O&M) costs, the need for major equipment upgrades or replacements, and future fuel and pollution (including potential carbon dioxide) costs. The LSEs also considered the impact of increasingly stringent Environmental Protection Agency regulations on the costs and operation of existing assets.

The assessment of new generation resources considered future fuel costs and environmental regulations. Fuel price forecasts were handled in different ways across studies; frequently NYMEX futures and proprietary fundamental energy market models managed by third parties were used in the planning studies. Additional detail on various fuel price forecasts and the treatment of uncertainty in planning models for LSEs in the western U.S. is provided by Larsen and Wilkerson (2012). In addition, LSEs estimated the capital cost of conventional and renewable generating options. Some also evaluated the availability and cost of procuring energy from wholesale power markets in future years.

For renewables, LSEs considered RPS requirements and the dependence of RPS requirements on future demand. In some cases RPSs specify a carve-out for solar generation or distributed generation (DG) that the LSEs also considered in designing feasible candidate portfolios. In Arizona, for example, Arizona Public Service (APS) and Tucson Electric Power (TEP) expected a significant portion of the DG requirement in the state's RPS to be met with distributed PV. Many included an assessment of future availability and timing of tax-based incentives for renewables and the resulting cost implications. Finally, many LSEs considered the ability to purchase unbundled renewable energy certificates (RECs) in lieu of RECs that are bundled with power that is delivered to the LSE. When there was no RPS requirement, some LSEs considered potential future prices for RECs from their renewable generation that they could sell to other markets.

A number of LSEs included an evaluation of existing and/or new transmission in their assessment of needs. In some cases, the need for transmission expansion was linked to particular generation options or import limitations that required reliance on broader wholesale markets to meet future needs. In other cases, transmission was driven by factors not linked to specific generation choices. Overall, LSEs did not appear to emphasize how transmission costs would vary across resource choices as much as other factors like resource capital and fuel costs. As noted by Schwartz (2012), availability of existing transmission is often the primary consideration for determining potential resources, rather than the cost of building new transmission, for LSEs in the western United States.

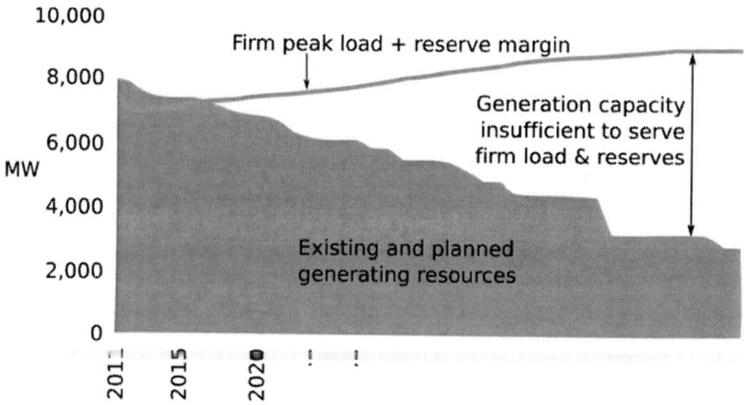

Figure 2. Example of expected future peak loads and existing resources (adapted from PSCo).

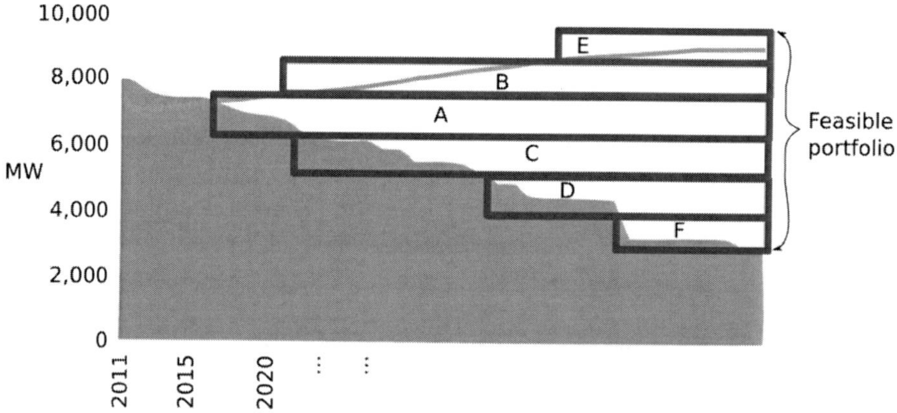

Note: The rectangles labeled A through F represent different resource options that illustrate one feasible portfolio (of many) to meet the peak load and planning reserve needs.

Figure 3. Example of the creation of a feasible candidate portfolio (adapted from PSCo).

3.2. Step 2: Creation of Candidate Portfolios That Satisfy These Needs and Constraints

The LSEs often needed to reduce a wide range of future options into a set of ***multiple*** feasible resource portfolios that could then be analyzed with available tools and methods. Generally, "feasible" meant that the candidate portfolios met all state and federal regulatory requirements. In almost all cases this also meant that the generating resources in each candidate portfolio could meet peak load (plus planning reserve margins) and annual energy demand (Figure 3).[10,11] A key element in this step was estimating the load-carrying capability of solar, i.e., the amount of additional load that can be met without decreasing the reliability of the system. LSEs used different approaches to estimating solar load-carrying capability. We refer to the LSE's estimate of load-carrying capability for each solar technology per unit of nameplate capacity as the "capacity credit" for that technology.

We found that three basic methods were used in current LSE planning studies to create feasible candidate portfolios (although some LSEs did not describe the approach they used):

- Manual/engineering judgment
- Commercial capacity-expansion models
- Ranking of wide range of resources

With the manual approach, no clear formula or objective was used to create feasible portfolios. Instead, logical arguments based on the trends observed in the assessment of needs and resources (engineering judgment) or stakeholder inputs were used to create feasible portfolios. Often a number of "bookend" portfolios were created to demonstrate the extent to which different portfolios lead to different revenue requirements or different levels of exposure to risk.

Commercial capacity-expansion models used mathematical search algorithms (typically based on linear, mixed-integer, or dynamic programming) to evaluate hundreds or thousands of potential resource combinations under a specific set of assumptions. The models selected the portfolio that minimized the present value of the revenue requirement (PVRR, including capital and dispatch costs). In some cases LSEs created multiple candidate portfolios by varying the assumptions input to the capacity-expansion model and selecting the optimal portfolio for each particular set of assumptions.

A few studies applied a ranking, often based on economic criteria, to the variety of resource options available to the LSE. In one case, non-economic factors (shortest time to bring renewables online and lowest environmental impact) were used to determine the composition of renewable resources included in candidate portfolios. Only the highest-ranked resources were then chosen for further evaluation in candidate portfolios.

3.3. Step 3: Evaluation of Candidate Portfolio Cost and Impacts

The costs and impacts of candidate portfolios were then quantitatively evaluated. Fixed capital, variable fuel, and O&M costs were estimated to calculate the PVRR of each portfolio. The PVRR is based primarily on the capital cost of each portfolio and the variable cost of dispatching each portfolio to maintain a balance between supply and demand over the planning period. The exposure of the portfolio to changes in costs based on uncertainties about the future was often evaluated to identify the relative risk of each option (Figure 4).

In this step, including solar in a portfolio resulted in changes in the dispatch of other generation resources, along with reductions in overall portfolio exposure to changes in future fuel and carbon prices. The dispatch of the candidate portfolio was typically simulated using a production cost model. The capital and fixed costs were commonly evaluated using a financial analysis model that transformed streams of future expenditures into a present value.

Portfolios were evaluated across LSE planning studies in three primary ways:

- **Deterministic evaluation:** A small number of LSEs simply estimated the PVRR for each candidate portfolio using one set of assumptions regarding future costs and needs. This approach is used in the illustration summarized later (Table 2).
- **Deterministic evaluation with specific sensitivity cases:** Many LSEs first estimated the PVRR of each portfolio based on a reference set of assumptions about future costs and needs and then examined the sensitivity of the PVRR to changes in those

assumptions using specific sensitivity cases. The sensitivity cases not only showed how the PVRR of each portfolio was exposed to changes in assumptions, but also how the relative ranking of portfolios on a PVRR basis would change under different assumptions, including natural gas and carbon dioxide prices, load growth or energy efficiency program effectiveness, and retirement decisions for existing plants (often older coal plants). The uncertainties included in the sensitivity cases were often justified by the analysts' assessments of future resource options and costs in Step 1. Typically the sensitivity cases changed one variable at a time, with a high and low case for each variable. In other cases multiple variables were changed simultaneously (e.g., high gas price and high carbon dioxide price vs. low gas price and low carbon dioxide price) to examine "best case" or "worst case" scenarios. The high and low variable estimates were often based on values that showed the upper or lower range of what was considered reasonably possible.[12]

- **Risk analysis using Monte-Carlo simulation:** A number of studies used Monte-Carlo simulation methods that allow many variables to change simultaneously within a distribution of potential future values. Monte-Carlo analysis required analysts to specify distributions of future values for each variable as well as the correlation between different variables. The PVRR of a portfolio was then estimated for hundreds of random draws from these distributions. As a result, the PVRR of each portfolio would also be a random variable with its own distribution of potential outcomes. Often the LSEs summarized the distribution of the PVRR for each portfolio with the average over all draws (the expected PVRR) and some metric that described the upper tails of the distribution (the risk of the portfolio).[13] Variables used in the Monte-Carlo simulation often included natural gas prices, load variations, and carbon dioxide prices. Again, the choice of these variables was often justified by the analysts' assessments of future resource options and costs in Step 1. In some cases, hydropower availability, generating plant forced outages, and wholesale electricity prices (among other variables) were included in the Monte-Carlo analysis. Although Monte-Carlo methods can account for correlation between uncertain variables, LSEs often assumed that all variables were uncorrelated.

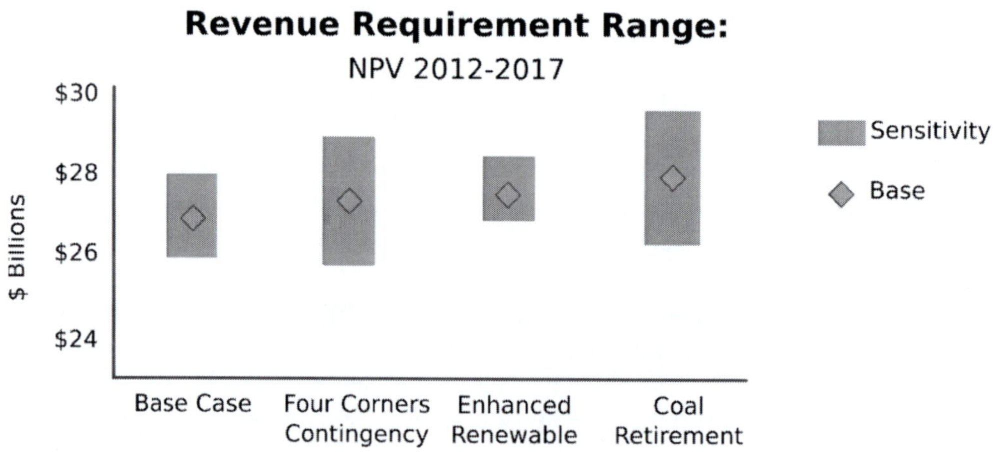

Figure 4. Example evaluation of the sensitivity of revenue requirement to different assumptions about the future for various portfolios (adapted from APS).

3.4. Step 4: Selection of Preferred Portfolio

After the analysis in step 3, four main methods were used to identify the preferred portfolio among the multiple candidate portfolios:

- **Lowest PVRR:** In a number of cases, the LSE's preferred portfolio was the portfolio with the lowest PVRR. Some LSEs did not assess the risk or degree to which the PVRR of different portfolio options would change if assumptions about the future changed.
- **Qualitative tradeoff between low PVRR and low risk:** Where scenario analysis was used to evaluate how the PVRR might change with different assumptions about the future, LSEs would sometimes use those results to adjust their preferred portfolio in an ad-hoc fashion to be a portfolio that was relatively low cost but also less risky. In some cases, for instance, the portfolio with the lowest expected cost might rely on purchases from the wholesale market instead of the LSE building or contracting for a new asset. The PVRR of this portfolio could increase greatly if wholesale prices varied within the plausible range posited by the LSE. In that case the LSE would choose a portfolio that had the next lowest cost but less exposure to the risk of high wholesale power prices.
- **Detailed weighting between expected cost and value at risk:** Some LSEs weighted the portfolio's expected PVRR against a quantitative measure of the portfolio's risk, such as the value at risk (e.g., the PVRR of the portfolio at the 90th percentile in the Monte-Carlo analysis). To estimate the expected PVRR and the PVRR at risk, the LSE had to perform a Monte-Carlo analysis with several hundreds of different potential futures.
- **Detailed weighting among many factors:** In some cases, additional factors were used to evaluate the attractiveness of each portfolio, e.g., by scoring based on cost, risk, reliability, resource diversity, or other factors.[14] This approach expands on the cost vs. risk tradeoff by adding other factors that may be important but are not clearly covered in the quantitative portfolio evaluation.

3.5. Step 5: Resource Procurement

Often LSEs created IRPs using publicly available information and generic representations of generating options. When procuring resources, the LSEs solicited bids to meet the needs identified in the preferred strategy. These bids could differ from the assumptions in the planning study in terms of cost, resource type, resource location, and generating profile. LSEs that also owned generation assets (like many western U.S. utilities) could also propose to build new generation that would satisfy the needs identified in the planning study (i.e., the "build" option rather than the "buy" option). Some LSEs used an independent evaluator to compare bids in order to mitigate potential conflict of interest. In most cases, the LSE's RFP described how the economic attractiveness of each bid would be estimated along with the information required by the LSE or the independent evaluator to evaluate each bid. LSEs often included non-economic factors such as bidder experience as part of the bid evaluation.

3.6. Bringing It All Together: A Simple Illustration of a Resource Planning Study with PV

To illustrate the basic mechanics of many of the planning studies that included solar, we present a simple conceptual example of how different resource portfolio options can be created and compared. We then use this simple example to illustrate the economic value of solar and the key drivers of this value.

3.6.1. Illustrative feasible candidate portfolios

In this simple example, we consider a hypothetical LSE with an expected peak load and planning reserve margin of 10,000 MW and electricity demand of 28.3 TWh in a future year (Figure 5).

To satisfy this future demand, we create the three feasible candidate portfolios shown in Figure 5. Each portfolio maintains all of the existing generation capacity; the portfolios differ only by the generation capacity added to satisfy future needs. Portfolio 1 adds only new combustion turbine (CT) capacity, Portfolio 2 adds CT and combined cycle gas turbine (CCGT) capacity, and Portfolio 3 adds CT, CCGT, and PV (500 MW) capacity. In each case the total capacity of all resources is sufficient to maintain the same level of reliability across all portfolios.

Assuming in this example that PV has a *capacity credit* of 50% of its nameplate capacity for this particular LSE and portfolio, then only 250 MW of the 500 MW of PV count toward the peak load and planning reserve target (the actual estimates of the capacity credit used by LSEs and the methods used to estimate the capacity credit are described in the next section). The addition of 500 MW of PV to Portfolio 3 therefore allows 250 MW of capacity (CT in this case) to be removed relative to a similar portfolio without PV (e.g., Portfolio 2). Since adding 500 MW of PV only removes 250 MW of CT, the overall nameplate capacity of Portfolio 3 is 250 MW greater than the overall nameplate capacity of Portfolio 1 or 2. The annualized fixed cost of each portfolio (excluding the capital cost of PV) is shown in Row "d" in Table 2.

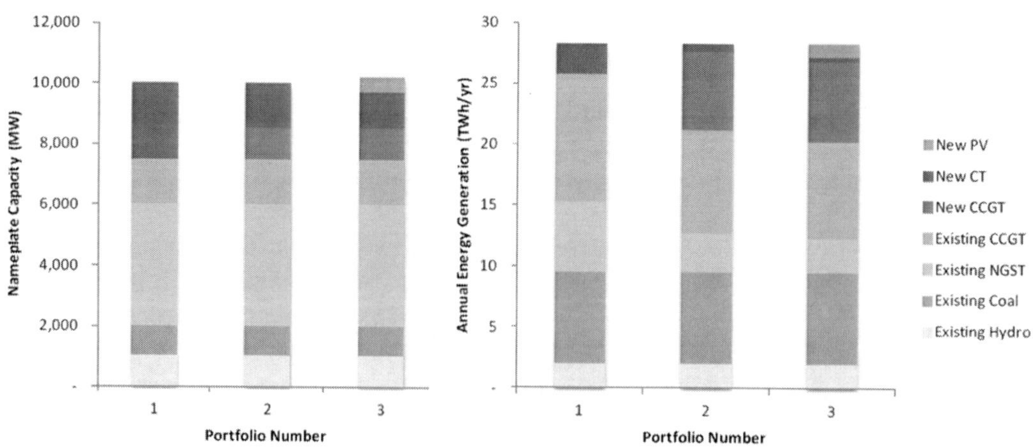

Figure 5. Portfolio composition and hypothetical dispatch.

Table 2. Evaluation and comparison of three hypothetical portfolios

		Units	Portfolio 1: Only Capacity	Portfolio 2: Capacity and Energy	Portfolio 3: Capacity, Energy, and PV	Notes and assumptions:
	Peak Load + Reserve Margin	(MW)	10,000	10,000	10,000	
	Existing NGST	(MW)	4,000	4,000	4,000	$30/kW-yr fixed cost
	Existing Coal	(MW)	1,000	1,000	1,000	$50/kW-yr fixed cost
	Existing CCGT	(MW)	1,500	1,500	1,500	$30/kW-yr fixed cost
	Existing Hydro	(MW)	1,000	1,000	1,000	$50/kW-yr fixed cost
	New CT	(MW)	2,500	1,500	1,250	$150/kW-yr fixed cost
	New CCGT	(MW)	-	1,000	1,000	$200/kW-yr fixed cost
	New PV	(MW)	-	-	500	$250/kW-yr fixed cost after ITC, 50% capacity credit
	Total Nameplate Capacity	(MW)	10,000	10,000	10,250	
	Annual Demand	(TWh/yr)	28.3	28.3	28.3	
	Existing NGST	(TWh/yr)	5.8	3.2	2.8	$60/MWh variable cost
	Existing Coal	(TWh/yr)	7.5	7.5	7.5	$25/MWh variable cost
	Existing CCGT	(TWh/yr)	10.5	8.5	8.0	$50/MWh variable cost
	Existing Hydro	(TWh/yr)	2.0	2.0	2.0	$5/MWh variable cost
	New CT	(TWh/yr)	2.50	0.60	0.42	$80/MWh variable cost
	New CCGT	(TWh/yr)	-	6.5	6.5	$40/MWh variable cost
a	New PV	(TWh/yr)	-	-	1.10	25% capacity factor
	Total Annual Generation	(TWh/yr)	28.3	28.3	28.3	
b	Existing Fixed Cost	($M/yr)	$ 265	$ 265	$ 265	
c	New Fixed Cost (Excluding PV)	($M/yr)	$ 375	$ 425	$ 388	
d = b + c	Total Fixed Cost (Excluding PV)	($M/yr)	$ 640	$ 690	$ 653	
	Variable Cost (with no PV integration cost)	($M/yr)	$ 1,261	$ 1,113	$ 1,048	
e	Variable Cost (Including PV integration cost)	($M/yr)	$ 1,261	$ 1,113	$ 1,053	Assumed $5/MWh integration cost for PV
f = d + e	Fixed + Variable Cost (w/o PV capital cost)	($M/yr)	$ 1,901	$ 1,803	$ 1,706	
g	PV Fixed Cost	($M/yr)			$ 125	
h = f + g	Total Cost (incl. PV Cost)	($M/yr)	$ 1,901	**$ 1,803**	$ 1,831	
i = f2 - f3	Avoided Cost of PV	($M/yr)			$ 97	
j = g - i = h3 - h2	Net Cost of PV	($M/yr)			$ 28	
k = g / a	PV Cost	($/MWh)			$ 114	
l = i / a	**Avoided Cost of PV**	**($/MWh)**			**$ 88**	
m = j / a = k - l	Net Cost of PV	($/MWh)			$ 26	

3.6.2. Evaluation of candidate resource portfolios

The dispatch of the existing and new generation, typically estimated using a production cost model, is assumed in this example to differ based on the resources available in each of the portfolios, as shown in Figure 5. Portfolio 1 requires significantly more of its energy to be generated by CTs and existing natural gas steam turbines (NGSTs) compared to Portfolio 2 and 3, which both have more CCGTs and PV. In particular, the addition of PV in Portfolio 3 displaces energy from new CTs, existing NGSTs, and existing CCGTs. We assume hydropower is dispatched differently in each portfolio, although the annual energy production remains the same. Owing to the relatively low variable cost of coal, the addition of PV to the portfolio does not affect the dispatch of the existing coal. In addition, PV is assumed to contribute an additional $5/MWh *integration cost* due to factors not represented in the other variable cost estimates such as the cost of day-ahead forecast errors and/or additional ancillary services (the actual integration costs used by LSEs and the methods used to estimate them are described in the next section). Including this integration cost, the annualized variable cost of each portfolio based on the assumed dispatch is shown in Row "e" in Table 2. The total annualized cost of each portfolio (both fixed and variable and including the capital cost of PV) is shown in Row "h."

3.6.3. Comparison of portfolios and selection of preferred portfolio

In this simple example, Portfolio 2 is the least-cost portfolio, as shown by Row "h" in Table 2, with an annualized cost of $1,803 million/yr. This portfolio can serve as a reference point to understand two key concepts regarding the relative economic value of PV: the *avoided cost* of PV and the *net cost* of PV.

The avoided cost of PV shows how much would be saved if the energy from the 500-MW PV plant could be added to Portfolio 3 without any capital cost associated with the PV installations. The cost of Portfolio 3 with no PV cost would be $1,706 million/yr (Row "f"). The avoided cost of the energy from PV relative to the preferred portfolio, Portfolio 2, would be $97 million/yr (Row "i") or $88/MWh (Row "l").

The capital cost of PV is not free, however, and in this case has an annualized cost of $125 million/yr for 500 MW after reducing the cost by the current Investment Tax Credit (or a levelized cost of $114/MWh[15]). Factoring this cost into Portfolio 3 pushes its cost above the cost of Portfolio 2. The net cost of PV is the capital cost of PV ($125 million/yr) less the avoided cost of PV ($97 million/yr). In this case the net cost of PV is $28 million/yr (Row "j") or $26/MWh (Row "m"). Based on how this cost was derived, the net cost of PV ($28 million/yr) can be added to the total cost of the least-cost portfolio (Portfolio 2: $1,803 million/yr) to come up with the total cost of Portfolio 3 including the cost of PV ($1,831 million/yr). In other words, the net cost of PV in this case represents the total cost above the cost of the least-cost portfolio due to the addition of PV to Portfolio 3. Further, if the avoided cost of PV were to increase by $26/MWh (due to increased prices for fossil fuel or carbon dioxide for instance) or the capital cost of PV were to decrease by $26/MWh, then the net cost of PV would be $0/MWh. In this case the total cost of Portfolio 3 would be equivalent to the total cost of the current least-cost portfolio.

3.6.4. Overall economic assessment of PV

This simple example highlights a few important points about this approach:

- **The composition of each portfolio impacts the apparent avoided costs of PV.** We could have just as easily added PV only to the much more expensive Portfolio 1. The apparent avoided costs of PV would have dropped tremendously in this case owing to the high operating cost of the CTs in the first place. To gauge the true avoided cost of PV, and hence the relative cost of any portfolio that includes PV, the portfolios must be well designed to minimize costs. Another way to ensure this would be to create and evaluate many combinations of portfolios so the "right" combination that accurately assesses the value of PV is included in the set of examined portfolios. Commercial capacity- expansion models, used by a number of the LSEs to identify feasible candidate portfolios, generally evaluate hundreds of combinations of different generation resources while searching for portfolios that minimize the PVRR. While that process was not replicated in this simple example, these tools can help identify good combinations of resources for candidate portfolios. For instance, a capacity-expansion model could identify a reference portfolio that is used to evaluate marginal changes in costs and benefits with marginal changes in the composition of that portfolio.

- **It is important to create a new portfolio with PV rather than simply adding PV to the original least-cost portfolio.** In the example, Portfolio 3 has 250 MW less CT capacity than Portfolio 2 owing to the 50% capacity credit of PV. Had we simply added PV to Portfolio 2 without accounting for the capacity credit of PV, the avoided cost would drop from $88/MWh to $54/MWh. The difference in avoided cost with and without consideration of the capacity credit of PV can be considered the *capacity value of PV*, which in this case is responsible for $34/MWh of the total avoided cost of PV. This shows that it is not appropriate simply to add PV to other portfolios already designed to satisfy capacity needs. The addition of new resources to a portfolio should be accompanied by a removal of resources that are no longer needed to maintain the same level of reliability across portfolios.
- **The dispatch of generation in each portfolio also contributes to the avoided costs.** In addition to the capacity credit of PV, the remaining contributor to the avoided cost is the avoided variable costs of the conventional power plants less the integration costs. The *energy value of PV*, or the avoided cost prior to accounting for the integration cost, in this case would be $59/MWh. Factors that impact overall production costs but are not accounted for in the dispatch models used to estimate the energy value (like day-ahead forecast errors or ancillary service [AS] costs) may require adjusting the estimate of the energy value. With an assumed $5/MWh integration cost for PV, for instance, the total avoided variable cost of PV would equal $54/MWh. This energy value will depend on the dispatch of all of the generation including the contribution of PV and thus highlights the importance of considering dispatch.

In summary, this simple example illustrates the creation of feasible portfolios, evaluation of the cost of each portfolio, and identification of the least-cost portfolio. It also introduces the concepts of *avoided cost of PV* (the difference in cost between the least-cost portfolio and the portfolio with PV assuming PV is free), *net cost* (the fixed cost of PV less the avoided cost), *capacity value* (the portion of the avoided cost due to the load-carrying capability of PV as represented by the capacity credit estimated by the LSE), *energy value* (the portion of the avoided cost from reducing the dispatch of conventional plants), and *integration cost* (an adjustment to the energy value due to factors like day-ahead forecast errors or AS requirements that are not captured in the portfolio evaluation). Throughout the remainder of this report we refer to these concepts to compare the methods used in practice by LSEs.

4. Analysis of LSE Treatment of Solar in Planning and Procurement

In this section we compare the treatment of solar across the various approaches used by the LSEs as each loosely followed the steps described in the previous section. Where appropriate, we also compare assumptions or approaches used in the LSE studies to the broader literature on solar valuation. We summarize key components included in the LSEs' evaluation such as the types of solar technologies considered; the treatment of the capacity value (with particular on the assumed capacity credit of the various solar technologies), energy value, and integration costs of solar energy; the LSEs' treatment of other factors including the risk reduction value of solar, impacts to the transmission and distribution system, and options that might mitigate solar variability and uncertainty; the methods LSEs use to design candidate portfolios of resources for evaluation within the studies; and the approaches LSEs use to evaluate the economic attractiveness of bids during procurement.

4.1. Solar Technologies Considered in Planning and Procurement

The potential solar technologies assessed by LSEs varied across planning studies. A list of solar technologies included in at least one of the studies is presented in Table 3.

A large range of solar technologies has been demonstrated commercially (Arvizu et al. 2011). LSEs might want to consider, at least at a screening level, this large range in their planning studies. Flat-panel PV technologies (both fixed and tracking) are suitable for much of the United States, whereas parabolic-trough and power-tower CSP with or without thermal storage or natural gas augmentation are suited for regions with ample direct normal insolation.[16] These technologies are all considered mature enough for commercial application (EASAC 2011). In addition, several solar thermal gas hybrid plants, each with more than 20 MW from solar, are now in operation or under construction around the world (Arvizu et al. 2011). Other technologies, such as solar chimneys, are still in the pilot or early demonstration stage; the lack of detailed cost and performance data for these technologies makes them more difficult to assess than PV or CSP.

Table 3. Solar technologies included in assessment of potential future resources

Solar technology category	Variation	Integrated thermal storage	Natural gas firing in boiler
Photovoltaic	Fixed	N/A	N/A
	Single-axis tracking	N/A	N/A
	With lead acid battery	N/A	N/A
Concentrating solar power	Parabolic trough	None	No
	Parabolic trough	None	Yes
	Parabolic trough	3 hours	No
	Parabolic trough	6-8 hours	No
	Solar power tower	7 hours	No
	Solar chimney (or solar updraft tower)	None	No
Solar thermal gas hybrid plants (or integrated solar combined cycle, ISCC)		N/A	N/A

4.2. Recognition of Solar Capacity Value in Planning Studies

4.2.1. Creation of feasible candidate portfolios implicitly provides capacity value

The capacity value of solar is based on the reduced need for new conventional capacity to maintain reliability. As shown in Section 3.6, adding solar to a portfolio can displace new CT or CCGT capacity in that portfolio. In almost all LSE planning studies reviewed here, the amount of resources added to each portfolio (including solar) was sufficient to meet the LSE's forecasted peak load and planning reserve margin over the planning horizon. These portfolios accounted for the lower load-carrying capability of wind and some solar technologies (see Section 4.3). Similarly, we rarely found cases in which the portfolio capacity exceeded the peak load and planning reserve margin.[17] As a result, adding solar reduced the need for some other capacity resource to meet the peak load and planning reserve margin (or maintain constant reliability across portfolios). In a few cases, the LSE's modeling approach reduced wholesale power market exposure during peak-load periods by adding solar; the LSE assumed the price of power during peak-load periods would include the long-run cost of building a new CT. The displacement of capital and/or variable cost of those capacity resources therefore affected the estimate of the PVRR of the portfolios with solar. The capacity value of the solar technologies was then embedded within the estimate of the PVRR as a combination of the capacity credit of solar and the avoided cost of the capacity resource displaced by the load-carrying capability of solar.

4.2.2. Magnitude of capacity value is driven partly by capacity cost and timing of capacity need

We found that the magnitude of capacity value was driven in part by the assumed cost of capacity and timing of capacity need. In many cases the capacity resource displaced by the load- carrying capability of solar appeared to be the full fixed cost of a CT (e.g., the capacity value of 100 MW of solar with a load-carrying capability of 50 MW would be based on the full fixed cost of 50 MW of new CTs). In those cases the LSE's assumed cost of CTs impacted the capacity value of solar. In a few cases the timing of capacity need also impacted the capacity value of solar: the capacity value could be lower if excess capacity was available in early years. For example, if an LSE did not need additional capacity to meet its peak-load and planning reserve margin for another 5 years, then adding solar now to meet an RPS requirement did not offset the need for other capacity for the first 5 years. The overall capacity value of that solar plant would therefore be lower than if the LSE currently had a need for new capacity.

4.2.3. Lumpiness of capacity options can affect the capacity value of solar

In at least one case (Tri-State), the alternative conventional generation options were very "lumpy" (the only individual conventional power plant options were 290 MW or larger), resulting in no change in conventional generation expansion when small amounts of solar were added. Tri-State did not appear to adjust for this lumpiness, e.g., they did not allow for off- system sales of excess capacity. This lumpiness of conventional capacity options may therefore result in no apparent capacity value of solar even though solar contributes to resource adequacy requirements. Including smaller investment options such as 50-MW or smaller CTs could minimize this issue. These smaller CTs were often included in other LSE studies. Alternatively, LSEs could avoid this problem by always including the option to sell excess capacity or purchase any deficit in capacity in any year at the price of the fixed cost of a peaker plant. More realistic representation of the lumpiness of capacity could be included by restricting those purchases and sales of capacity to increments similar to the size of a peaker plant (e.g., a 50-MW CT).

4.3. Estimates of Solar Capacity Credit in Planning Studies and Broader Literature

The amount of other generation capacity that could be displaced by solar while still meeting the peak load and planning reserve requirement depended on the load-carrying capability of the solar technologies assumed by LSE planners. The capacity value of solar (in economic terms) was therefore greatly impacted by the capacity credit the LSE assigned to solar, which was affected in part by the method the LSE chose for estimating the solar capacity credit. Across all LSEs, the solar capacity credit was always positive (with one exception for small PV in California[18]), even though many solar technologies cannot generate at night, and the daytime output varies with cloud cover.

The methods used in estimating the capacity credit of solar varied widely across LSEs, from detailed loss of load probability (LOLP) studies (e.g., APS, Public Service of Colorado [PSCo]) to estimates of solar generation during peak-load periods (e.g., Public Service of New Mexico [PNM], TEP, the California Investor-Owned Utility [CA IOU] process,[19] NV Energy, Idaho Power, Tri-State) to rules of thumb based on engineering judgment (e.g., Portland General Electric [PGE]). A number of utilities did not specify the method used to estimate the capacity credit of solar, e.g., Los Angeles Department of Water and Power (LADWP), Imperial Irrigation District (IID), and SRP.

4.3.1. Factors impacting solar capacity credit at low solar penetration

Aside from the approach used to estimate the capacity credit, the correlation of load and solar generation and the solar technology impacted the capacity credit of solar (Table 4, Figure 6). As expected, the capacity credit assigned to solar in regions with good coincidence of insolation and load (e.g., summer afternoon peaking loads) used a higher capacity credit than regions where the coincidence is lower (e.g., winter night peaking loads). Even in regions with summer afternoon peaking load, however, the capacity credit of PV or CSP without thermal storage is not expected to be 100% since peak solar production often occurs earlier in the day (1 pm or earlier) than peak loads (often after 2 pm). The resulting capacity credit applied to PV in the creation of portfolios ranged from 27% to 77% in regions with relatively good coincidence of PV generation and peak loads to 5% for an LSE in the Pacific Northwest where peak loads occur during winter nights (e.g., PGE[20]). In some cases separate capacity credits were estimated for single-axis tracking and fixed PV; the capacity credit for tracking PV appeared to be higher. This higher capacity credit for tracking PV is supported by a detailed comparison of the capacity credit for various PV configurations by Madaeni et al. (2012c). They found that the capacity credit of PV in the Western Electricity Coordinating Council (WECC; assuming zero PV penetration and no transmission constraints across

WECC) was 56%–75% for fixed PV, 70%–87% for single-axis tracking PV, and 71%–93% for double-axis tracking PV, depending on location (Figure 7).

The capacity credits used by LSEs for CSP with multiple hours of thermal storage or natural gas augmentation (87%–100% of nameplate capacity) are higher than the capacity credits for CSP without thermal storage or natural gas augmentation[21] (55%–87% of nameplate capacity).

Table 4. Capacity credits applied by LSEs in planning studies

Technology	Sub-category	Capacity credit range	LSEs within range
PV	Excluding Pacific Northwest	27% –77%	APS, CA IOU process, Duke Energy, LADWP, NV Energy, PNM, PSCo, TEP
	In Pacific Northwest	5% –36%	Idaho Power, PGE
	With lead-acid battery	100%	PNM
CSP	Without thermal storage or natural gas augmentation	55%–87%	APS, CA IOU process, PNM, TEP
	With thermal storage or natural gas augmentation	87%–100%	APS, CA IOU process, Idaho Power, PSCo, TEP

Note: Imperial Irrigation District (IID) appears to have assumed a 100% capacity credit for PV and a solar chimney. This assumption is excluded from the table because it is not supported by detailed analysis from IID or elsewhere. The California IOU process assumed small-scale PV would have a 0% capacity credit in its net cost ranking; this was also excluded from the table. Tri-State indicated that it estimated the capacity credit of PV to range from 20% to 57%, but it does not specify what value was used in its study. This range is also excluded from the table. All of these excluded values are shown in the corresponding figure.

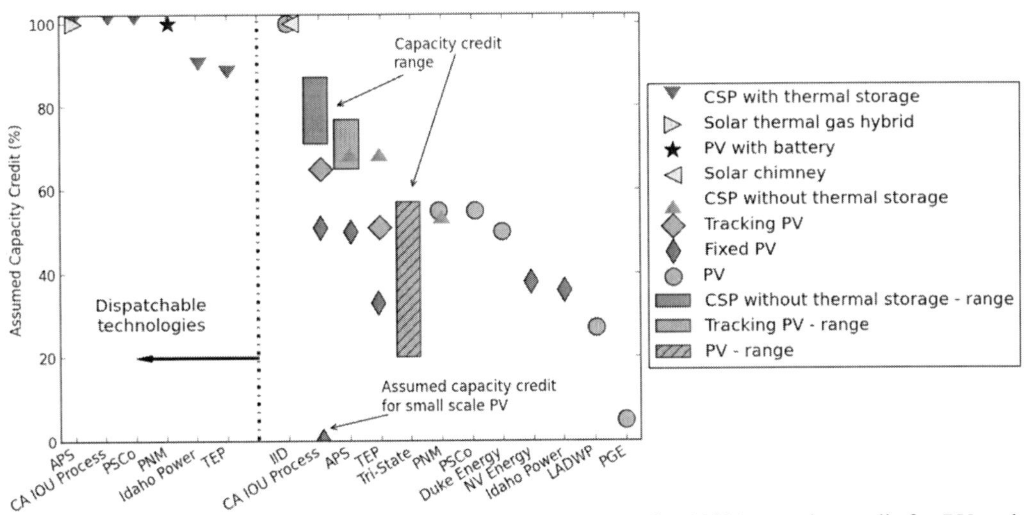

Note: Imperial Irrigation District (IID) appears to have assumed a 100% capacity credit for PV and a solar chimney.

Figure 6. Capacity credits applied by LSEs in planning studies.

Madaeni et al. (2012b) suggest that multiple hours of thermal storage are needed to obtain a capacity credit over 95%. The studies with CSP with thermal storage that describe the number of hours of storage capacity all have 3 hours of storage or more. The one LSE that considered a lead-acid battery coupled to a PV system, PNM, assumed a capacity credit of 100% for the combined resource. It is not clear how many hours of storage PNM assumed (its current demonstration PV/battery plant has 2 hours of storage capacity), but it is likely that multiple hours of storage would be required to justify such a high capacity credit.

The range of capacity credits used by LSEs in planning studies largely fall within the range reported in the broader literature for low-penetration PV and CSP (Figure 7) and the range reported using various time-based approximation methods within the United States (Rogers and Porter 2012). That said, the wide variation across the LSE studies, even for similar solar technologies in similar locations (e.g., the capacity credit for fixed PV used by APS, TEP, and NV Energy substantially differ even though all the LSEs are in the Southwest) suggests that variations in methods and tools used to estimate the capacity credit are responsible for a portion of the variation.

4.3.2. Reduction in capacity credit with increasing solar penetration levels

Only APS appeared to account for the potential change in the capacity credit of solar technologies with increasing penetration (see Appendix A for details on the APS PV capacity credit). LSEs considering large amounts of solar would ideally account for the potential change in the capacity credit of solar with increasing penetration. A number of LSEs are aware of this issue, but in some cases regulatory processes are used to establish the methods or assumptions used for estimating the capacity credit of different technologies. In those cases stakeholders and regulators should consider evaluating changes to existing practices if they do not account for changes in capacity credits with increasing penetration.

In contrast to assuming an unchanging solar capacity credit with penetration, detailed studies of the value of solar indicate that the marginal capacity credit of fixed PV, tracking PV, and CSP without thermal storage (or without natural gas augmentation) can even decline at relatively low penetration levels (Figure 8). While there is a wide range of capacity credits for PV at low penetration levels, the marginal capacity credit declines to 30% or lower at PV penetration levels above 10% on an annual energy basis across all of the examples in the literature. Analysis of many studies in the United States and Europe also shows that the capacity credit of wind generally decreases with increasing wind penetration (Gross et al. 2007, Holttinen et al. 2011).

The merits of CSP with thermal storage or natural gas augmentation increase relative to solar technologies without these features partly because they are able to maintain a higher capacity credit with increasing penetration. For example, Mills and Wiser (2012) found the marginal economic value of additional CSP with 6 hours of thermal storage to be about $35/MWh greater than that of single-axis tracking PV when either technology was at 10% penetration on an energy basis. If LSE planning studies do not account for changes in solar capacity credit with increasing penetration, they will not recognize this difference in value between various solar technologies.

The discrepancy between the capacity credit used in planning studies and the potentially lower actual capacity credit at high penetration may not be an important oversight in most current planning studies that only include low solar penetration cases. The studies by the CA IOUs, however, have scenarios that approach high penetration levels at which changes in the capacity credit could be important. APS also has scenarios in which PV penetration exceeds 5% on an energy basis, but the PV capacity credit used by APS accounted for effects of PV penetration across all scenarios. As the penetration of solar grows or as more LSEs consider portfolios with large shares of solar, it is important that changes in the capacity credit with increasing penetration be carefully considered.

It may also be possible to maintain a high capacity credit for PV with increasing penetration in one LSE's region by selling the output of PV during peak periods to neighboring LSEs that have lower PV penetration.[22] This possibility, however, was not considered in the studies. Part of the challenge for realizing this value in planning studies is the need for coordination between various LSEs. Even without increased amounts of solar (and wind), LSEs would benefit from coordinated evaluation of reliability with other LSEs in the same region. For example, Ibanez and Milligan (2012) find that the load that could be served reliably with aggregation at the sub- region level would be 14% greater than the load that could be served if each LSE considers only its own loads and resources in WECC. The fact that most LSEs consider only their own load and resources, even with the potential value from coordination, suggests that it may be challenging to change this existing practice as more solar is added. To help overcome the coordination challenge, with or without increasing amounts of solar or wind, regional reliability entities such as WECC could act as a repository for LSEs' planning assumptions and conduct regional adequacy evaluation studies.

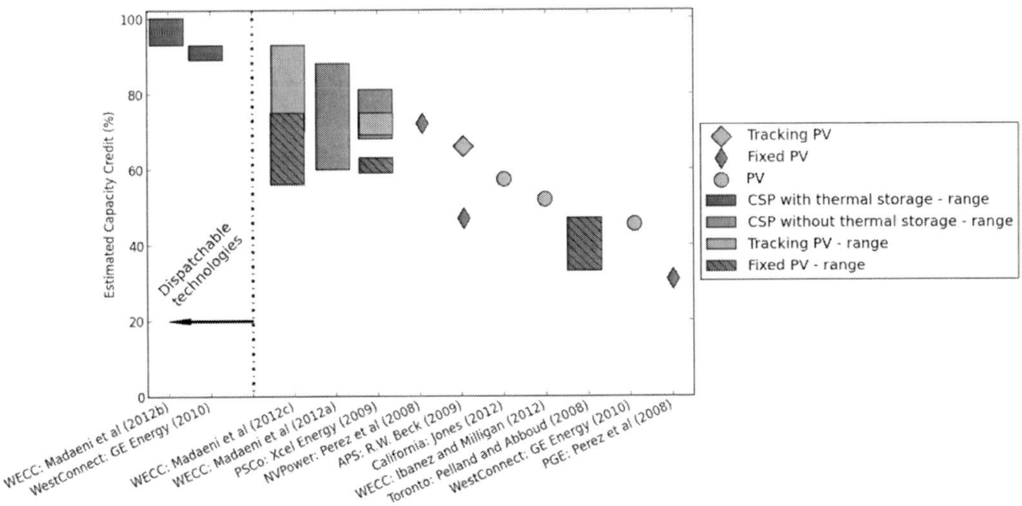

Note: Original PV capacity credit from GE Energy (2010) was reported based on DC nameplate capacity; here it is converted to AC nameplate capacity.

Figure 7. Estimated capacity credits at low penetration from studies that use LOLP-based methods.

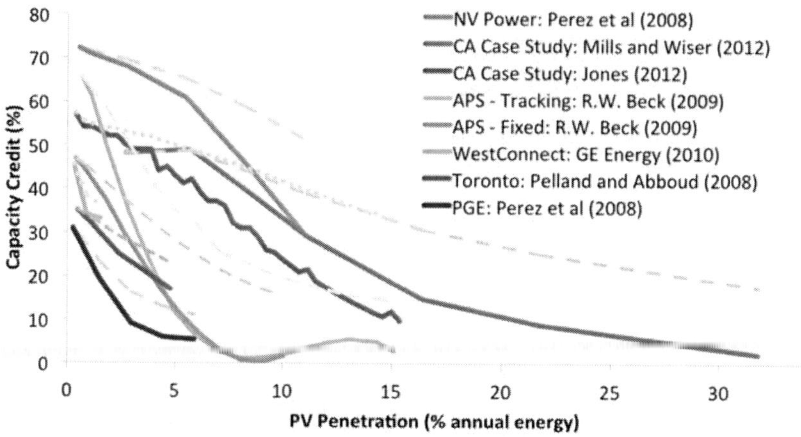

Notes:

- Mills and Wiser (2012) assume single-axis tracking with latitude tilt.
- GE Energy (2010) and Jones (2012) use PV profiles from a mixture of fixed and tracking PV.
- The scenarios in GE Energy (2010) with PV also have increasing penetrations of CSP with thermal storage and wind.
- In Perez et al. (2008), fixed PV with 30-degree tilt is assumed. Capacity credit is based on their estimate of the effective load-carrying capability (ELCC) of PV. Capacity penetration is converted to energy penetration assuming: NV Power load factor is 42% (based on NV Energy 2012 IRP), NV Power PV capacity factor is 23% (estimated from NREL Solar Advisor Model), PGE load factor is 58% (based on PGE 2009 IRP), and PGE PV capacity factor is 17% (based on PGE 2009 IRP).
- In Pelland and Abboud (2008), capacity penetration is converted to energy penetration assuming that Toronto's load factor is 55%. Fixed PV with 30 degree-tilt is assumed. We show only the results from a south-facing orientation.
- Capacity penetration used in R.W. Beck (2009) is converted to energy penetration assuming: APS load factor is 48% (based on APS 2012 IRP), APS tracking capacity factor is 30%, and APS fixed capacity factor is 23% (based on NREL Solar Advisor Model).

Figure 8. PV capacity credit estimates with increasing penetration levels (dashed line is average capacity credit, solid line is incremental capacity credit).

4.3.3. Planning studies should consider improving estimates of solar capacity credit

Although the range of capacity credits applied to PV technologies and CSP plants with thermal storage or natural gas augmentation is broadly consistent with the range identified in detailed studies at low solar penetration levels, relatively few LSEs used detailed studies to estimate the capacity credit. It appears that only PSCo and APS used LOLP-based reliability analysis of PV among the LSEs considered here. Most other capacity credit estimates were based on rules of thumb, evaluation of solar generation during peak-load periods, or assumptions. The numeric values appear plausible, but, owing to the importance of reliability in planning studies and the importance of the capacity credit for solar's economic value, detailed LSE-specific reliability assessments of capacity credit would be appropriate. In the Pacific Northwest, an area dominated by energy-constrained hydropower resources, these assessments should seek to account for energy limits in the reliability assessments. The importance of estimating solar load-carrying capability will grow as solar's costs decline and it becomes more competitive with other resource options.

Table 5. Temporal resolution of production cost models used in LSE planning studies

Temporal resolution of production cost model	Planning Studies	Production cost model (Company)
Hourly for a full year	CA IOU process	PLEXOS (Energy Exemplar)
	Idaho Power	AURORAxmp (EPIS)
	LADWP	PROSYM (Ventyx)
	NV Energy	PROMOD IV (Ventyx)
	PGE	AURORAxmp (EPIS)
	Tri-State	PROSYM (Ventyx)
Hourly for one week each month	PacifiCorp	PROSYM (Ventyx)
	PNM	PROVIEW (Ventyx)
	PSCo	PROVIEW (Ventyx)
Hourly for one day per month	APS	PROMOD IV (Ventyx)
On-peak and off-peak for each quarter	NPCC	N/A

4.4. Evaluation of the Energy Value of Solar Using Production Cost Models

4.4.1. Production cost models account for solar's ability to offset high cost fuels

Across all planning studies the variable costs associated with dispatching power plants to meet varying demand in future years was simulated using some form of production cost model. For candidate portfolios with solar, adding solar reduced fuel consumption and production costs associated with the dispatch of other power plants. The reduced production costs therefore reduced the variable cost component of the PVRR. These avoided production costs associated with the inclusion of solar in candidate portfolios are referred to as the energy value of solar.

Although the temporal resolution of the production cost models used in the planning studies varied (Table 5), most of the models should have high enough resolution to reflect correlations between solar generation and times when the fuel costs of conventional power plants are high. The planning studies should therefore reflect the relatively high energy value for solar found in the literature. Some planning studies used models with hourly time resolution over a full year. Other studies reduced the amount of computational processing by only selecting an hourly time series of data from one week each month. This approach maintained the hour-to-hour structure of the data within each month, while reducing the number of hours to process by about 75%. However, this approach introduces the possibility that the weeks chosen for analysis are missing periods with extreme events (e.g., potential solar curtailment on spring weekends with high solar and low load). APS further reduced the sampling to one day per month using hourly data. The lowest temporal resolution observed in evaluating portfolios was the use of average on-peak and off-peak values for each quarter of the year. This last approach was used in the Northwest Power and Conservation Council (NPCC) analysis, in which thousands of portfolios were evaluated over 750 different futures, necessitating a reduction in processing complexity. The system evaluated by the NPCC is hydro dominated, and energy deficits are a larger concern than transient events owing to the large nameplate capacity of the system and relatively low amount of energy storage. NPCC's focus on balancing energy over longer periods rather than balancing hour-to-hour supply and demand potentially justifies the use of lower temporal resolution. Such low temporal resolution, however, may not be as appropriate for regions dominated by thermal generation that are constrained primarily by capacity rather than by energy.

4.4.2. Production cost models account for changing energy value with increasing penetration

The production cost models used in planning studies account for the reduction in marginal fuel savings as solar penetration grows and the net load becomes lower (i.e., the depth of the dispatch stack). The methods used by most LSEs in the sample therefore account for how adding more solar displaces production from resources with lower and lower variable costs. Consequently, the methods currently used in most LSE planning studies capture changes in the energy value of solar with increasing penetration. However, the studies that use low temporal resolution and lack operational constraints may miss important impacts due to the addition of solar at high penetration levels. Previous detailed analysis shows that it may be important to include high temporal resolution and operational constraints in production cost models to capture the energy value of solar at higher penetration levels (e.g., Mills and Wiser 2012).

The degree to which production cost models included operational constraints on conventional generation (e.g., ramp-rates, minimum generation, start-up time, and start-up cost) differed by study. The CA IOU process, APS, Tri-State, and PGE, for example, used hourly time resolution with a detailed production cost model that included unit-by-unit commitment, ramp-rate limits, start-up time, and minimum generation limits (PLEXOS, PROMOD IV, PROSYM, and AURORAxmp, respectively). The PROVIEW model, on the other hand, uses probabilistic techniques for estimating production costs of different portfolios that will not capture unit-by- unit commitment and constraints. Additional details regarding the capability and limitations of some of these models are described by KEMA Consulting (2003).

Where it is not possible to improve models used in planning studies to capture effects associated with extreme events or operational constraints on conventional generation (for computational or cost reasons), an alternative may be to conduct a side analysis using more detailed models to identify "correction factors" that adjust production costs based on portfolio composition.

Integration cost studies, for example, often identify costs or impacts associated with increased penetration of variable generation that are not captured in the less detailed models used in planning studies. These integration costs are then used to adjust results from the less detailed models such that the costs associated with complex unit-commitment or AS issues are partially reflected in the evaluation of candidate portfolios. This approach can cover many factors not otherwise addressed in the models used in planning studies (see Section 4.5).

Another important consideration for LSEs, particularly as solar penetration increases, is the degree to which the broader wholesale market becomes a market for selling excess power and how much neighboring LSEs procure solar. The degree to which LSEs considered the wholesale power market outside of their own resources or long-term contracts to meet energy needs or as a buyer of power from the LSE's resources varied across studies. PGE clearly included the wholesale power market in the West in its analysis, using the long-term capacity-expansion model AURORAxmp (EPIS) to create a long-term capacity-expansion plan and dispatch for the Western Interconnection.[23] PGE then used this simulated Western wholesale market as an alternative source of energy or buyer of energy from its resources. With such an approach it would be possible to capture the benefit of selling power from solar generation to other markets during times when solar penetration is high for a particular LSE. It would also be possible to evaluate the impact of other LSEs simultaneously increasing solar penetration, which could restrict opportunities to sell excess power to neighboring regions. The CA IOU process and Idaho Power also included the broader wholesale power market by dispatching resources across the entire WECC footprint in their production cost models (although they did not use a long-term capacity-expansion model to create the portfolios for other LSEs outside of their respective regions).

Another method, used by PNM for example, of modeling the wholesale market outside of the LSE region was to use Monte-Carlo based sampling from a historical wholesale price distribution. This approach ensures that events that have historically caused extreme changes in wholesale prices—such as generator failures, exercise of market power, or unexpected weather patterns (factors challenging to capture in production cost models)—are reflected in the distribution of wholesale prices used to evaluate portfolios. However, this approach does not capture potential changes in wholesale price distributions and regional price differences based on evolving factors, such as the increasing role of variable generation in other LSEs' portfolios.

LSE planning studies can be improved by clearly stating whether the broader market is being included in the analysis and, if it is not included, stating why the LSE assumes the broader market is not available to buy and sell power. In cases with increasing amounts of solar generation, LSEs might consider how much solar might be used in other parts of the market too.

4.4.3. Planning studies provide little detail on how thermal energy storage dispatchability is captured in production cost models

The planning studies do not appear to include detailed analysis of the potential for CSP with thermal storage to shift solar production to times of most value for a particular scenario. Similarly, few appear to account for the ability of CSP with thermal storage (or natural gas augmentation) to provide ancillary services or reduce the impact of operational constraints from conventional generation.

It is relatively straightforward to treat CSP with thermal storage as having a fixed dispatch profile in a production cost model, assuming that the thermal storage is always dispatched according to typical utility load patterns (e.g., GE Energy 2010). This simple approach ignores the additional flexibility benefits of CSP with thermal storage, including the potential provision of ancillary services. In scenarios with high renewable penetration, the most valuable dispatch pattern of CSP with storage may change. Fully incorporating CSP dispatch into the production cost model can identify the most valuable dispatch patterns for CSP with storage. There are challenges with implementing this in practice since thermal storage is an energy-limited resource—the thermal storage can be "charged" only by the solar field, compared with pumped hydro, which can be "charged" at any time from the grid—and CSP plants must operate above a minimum generation threshold (Sioshansi and Denholm 2010, Brinkman et al. 2012). None of the LSE planning studies describe how the dispatch of CSP with storage is incorporated into the production cost models.

It is easier to incorporate CSP with natural gas augmentation into production cost models. The production of the CSP plant is limited to the available solar profile and the capacity of the power block when natural gas augmentation is not used. When needed, natural gas can be burned in the steam boiler with a very high heat rate. PacifiCorp, for example, uses a heat rate of 11,750 BTU/kWh in its model of CSP with natural gas augmentation. The high heat rate (indicating the relatively low efficiency of burning natural gas to raise steam) suggests that the role of natural gas augmentation is only to increase the capacity value of a CSP plant, not to substantially increase the annual energy output of the plant. Depending on the number of hours per year that are critical for maintaining system reliability, the amount of natural gas burned in a CSP plant with natural gas augmentation can be low while still increasing the capacity credit of the plant.

The relative importance of dispatching CSP with thermal storage is not well known, but initial research indicates there is a benefit (Sioshansi and Denholm 2010, GE Energy 2010, Madaeni et al. 2012b, Mills and Wiser 2012). To capture the full benefits, representation of the dispatchability of CSP with thermal storage may need to be improved in production cost models. Alternatively, LSEs can conduct side analyses that identify the magnitude of the potential benefits from flexibly dispatching CSP with storage relative to a static dispatch profile. These benefits could then be used to adjust the production cost estimates for portfolios with CSP with storage when only a static dispatch profile is used in the planning study (this adjustment would essentially be the opposite of an integration cost that is commonly applied to variable generation resources).

4.5. Adjusting the Energy Value to Account for Integration Costs

When evaluating portfolios with variable generation, a number of LSEs factored in an integration cost related to resources with increased variability and uncertainty. In some cases, this integration cost was an adjustment to the production cost model results based on an assessment of factors that could not be addressed in the model runs. The reasons the factors could not be addressed directly included limited temporal resolution of the production cost model and/or missing constraints from the model (e.g., day-ahead commitment based on imperfect forecasts or changes in operating reserve requirements for portfolios with more solar). In some cases the justification for integration costs was not stated. Milligan et al. (2011) suggest that, owing to the challenges and common misunderstandings associated with integration costs, such calculations should be conducted carefully, if at all.

4.5.1. Increasing ancillary service requirements in production cost models to account for short- term variability and uncertainty

One approach to accounting for integration costs was to include the need for increased AS requirements directly in the production cost modeling of portfolios with solar.[24] To do this, LSEs used estimates of AS requirements from more detailed studies (or other analysis within the LSE's broader planning study) and then modified the AS requirements in the production cost model for portfolios with solar (a similar process would be used for portfolios with wind). NV Energy, for instance, increased the operating reserve requirement in its production cost model for a portfolio that included solar. The increase in AS requirements was based on a separate detailed study that focused on integrating large amounts of PV into the LSE's system (Navigant Consulting et al. 2011). As part of the CA IOU process, the

California Independent System Operator (CAISO), in conjunction with the IOUs and many other stakeholders, estimated the AS requirements and need for additional capacity to meet the increased AS and ramping requirements across different portfolios with renewable energy to meet a 33% RPS by 2020. The increased AS requirements in the production cost modeling implicitly lead to an integration cost embedded in the overall production cost and therefore in the PVRR estimate for each portfolio.

4.5.2. Adding integration cost estimates to production cost results

In other studies, an integration cost for solar was simply added to the production costs for portfolios with solar (as in the Section 3.6 example). The integration costs added to production costs in portfolios with solar ranged from $2.5 to $10.9/MWh (Table 6).[25] These integration costs were often specified for particular solar technologies but were not evaluated down to the granularity of individual solar projects. Generally, integration costs were assumed higher for PV than for CSP with or without thermal storage, although Tri-State appears to assume the integration costs for CSP with 3 hours of thermal storage are the same as for PV (and wind).

Detailed estimates of solar integration costs tailored to address gaps in an individual LSE's production cost model were rare. Rather than conducting studies of solar integration costs, a number of LSEs pointed to other solar integration studies for cost estimates, or used wind integration cost estimates.

PSCo used a solar-specific integration cost study based on its system to estimate integration costs used in its planning study. The integration cost for solar in that study was based on the difference between the production costs over a year with and without day-ahead solar forecast errors. In the case with day-ahead forecast errors, the commitment of non-quick-start generation was based on the previous day's solar generation profile. The actual production costs were then based on dispatching the system with the actual solar generation profile for that day. No other adjustments were made to account for shorter-term variability or AS requirements in the PSCo integration cost study.

Table 6. Assumed integration costs used by LSEs to adjust production costs for portfolios with solar

Planning Studies	Integration Cost Added to Production Costs ($/MWh)			Notes
	PV	CSP without thermal storage	CSP with thermal storage	
PSCo	$5.15	N/A	$0	
APS	$2.5	$0	$0	
TEP	$4	$2	$0	
Tri-State	$5–$10	N/A	$5–$10	Most scenarios used low end of costs; scenarios with more renewables used higher costs
PGE	$6.35	N/A	N/A	
NPCC	$8.85–$10.9	N/A	$0	Integration costs assumed to 2024

4.5.3. Integration cost estimates would ideally be tailored to cover specific limitations in production cost models

Many LSEs could improve their estimates of the integration costs for solar. An LSE's definition of integration cost should ideally match with elements of the production cost model that are left out in order to avoid double counting or missing important components. In fact, the production cost model component of the LSE's planning study could be detailed enough to not need a separate integration cost estimate. With a sufficiently detailed production cost model and study methodology, the LSE could include all elements of variability and uncertainty in the production cost model runs while evaluating portfolios. The approach used by NV Energy and the CA IOU process is a good example: side analyses of AS requirements based on 1-minute load and solar profiles (and wind in California) were used to estimate the additional AS needs for portfolios with increased variable generation. These AS requirements were then directly included in the production cost model used to evaluate the performance of each portfolio.

This detail in analyzing each portfolio under each potential future, however, is still computationally challenging and may not yet be practical for many LSEs. In that case, the integration cost estimate used in the planning can be tailored to the limitations of the production cost models used by the LSE and the characteristics of the LSE's system (in terms of generation and demand-side flexibility, integration with other markets, and deployment of solar).

Given the current variation in LSE systems, tools, and methodologies used in planning studies, different approaches could generate different solar integration cost estimates. The importance of tailoring integration cost studies to fill gaps in the tools used by LSEs means no single integration cost estimate or methodology will necessarily be appropriate for all LSEs.

4.6. Additional Factors Included or Excluded from Planning Studies

4.6.1. Accounting for the risk-reduction benefits of solar in planning studies

The risk-reduction benefits of solar can be included in LSE planning assessments by accounting for uncertainty in future parameters—such as load growth, fossil fuel prices, and carbon policy— when evaluating candidate portfolios. Many of the planning studies accounted for the exposure of an LSE to changes in assumptions about the future when evaluating candidate portfolios, including portfolios with solar. The risk-reduction benefit of solar could be quantified as the reduced PVRR range for portfolios that include solar, as illustrated by the reduced sensitivity to uncertainties in the "Enhanced Renewable" portfolio in Figure 4. Although this risk-reduction benefit of solar was embedded in the results that LSEs used to identify preferred portfolios, none of the studies specifically isolated the degree to which solar alone reduces a portfolio's overall exposure to uncertainties.[26]

4.6.2. Unique costs or benefits of distributed PV in planning studies

Most LSEs did not distinguish between distributed PV and utility-scale PV or their respective benefits and costs. A few LSEs, however, either explicitly adjusted the PVRR to account for the presumed benefits of distributed PV or implicitly accounted for a portion of the benefits by distinguishing between solar sited near load centers and other forms of generation. Additionally, a few LSEs simply adjusted their peak load and energy forecasts based on estimates of future customer-sited PV. The CA IOU process explicitly included a distribution system cost-reduction benefit for portfolios with large amounts of distributed PV. The estimate of the distribution benefit applied to distributed PV varied by location but was most often around \$5/MWh (with a range of \$4.3 to \$26.2/MWh). An earlier analysis of the benefits of distributed PV conducted for APS (R.W. Beck 2009) found avoided distribution costs from distributed PV of \$0 to \$3.1/MWh, although these benefits were not explicitly mentioned in the APS planning study. In some studies, transmission expansion was required for portfolios with new distant resources, while less was required for portfolios with solar sited closer to load.

Distributed PV can also reduce transmission line losses. Production cost models with transmission network representation can account for reduced transmission losses when solar is located near major load centers (e.g., Lin 2012). The portfolios that included PV in major load zones would therefore reduce line losses in the production cost models, which would then reduce the PVRR for that portfolio, if the LSEs evaluated transmission line losses when evaluating portfolios. It is challenging to model line losses in production cost models, however, particularly when looking out many years to when the transmission network configuration may not be known. Furthermore, production cost models generally do not account for changes in losses in distribution systems, which means the impact of distributed PV on distribution line losses would be ignored in the studies that did not explicitly quantify these impacts. Although it is possible to estimate the impact on line losses of distributed PV (or the increase in losses from additional generation sited far from load centers), most planning studies did not appear to account for changes in line losses for different portfolios. If line losses are not accounted for directly in the production cost models, then approximation methods that account for the non-linear relationship between line loading and losses can be used to estimate the line-loss reduction benefit of distributed PV (e.g., Borenstein 2008, R.W. Beck 2009).

The transmission and distribution (T&D) impacts of distributed PV are complex and sensitive not only to penetration levels but also to a wide variety of localized factors (e.g., Shugar and Hoff 1993, Cossent et al. 2011, Katiraei and Aguero 2011). The cost impacts can be positive or negative. More LSE planning studies would ideally account for the costs and benefits of distributed PV on the T&D system as PV deployment increases and as tools to estimate these impacts improve.

4.6.3. Options to mitigate output variability and uncertainty of solar in planning studies

None of the planning studies in the sample explicitly identified synergistic benefits between PV and CSP or between solar and wind in portfolios.[27] Such benefits may have been implicitly recognized in production cost modeling where portfolios had combinations of renewable resources, but nothing was mentioned with respect to adjustments to integration costs or the capacity credit of technologies. The capacity credits for wind, CSP with thermal storage, and PV estimated in the Western Wind and Solar Integration Study showed a slight synergistic effect: the capacity credit for all three resources combined was 0.3 to 0.7 percentage points greater than the sum of the capacity credit of each resource alone (GE 2010). On the other hand, analysis of the capacity credit of wind and solar combinations in the Western Interconnection by Ibanez and Milligan (2012) appears to indicate a slight reduction (2.5 percentage points) in the capacity credit from the combination of 29 GW of wind and 14 GW of solar, enough to meet 8% and 3% of the annual WECC energy demand, respectively, relative to the capacity credit of each technology in isolation.

A number of LSE planning studies included several technologies that might be more attractive in scenarios with high solar penetration. Aside from including natural gas augmentation and thermal storage in CSP plants, for example, some LSEs included batteries with PV. PNM included a 1-MW PV system coupled with a lead-acid battery. PSCo added batteries to manually created renewable portfolios with PV. In contrast to a coupled battery and PV system, the operation of the battery in the PSCo portfolio was not coupled to the operation of the renewables; both the battery and renewables could be dispatched in a manner than minimized system costs. In the PNM and PSCo cases, any additional value from including the battery was not found to be sufficient to make up for its high cost. Other LSE studies included compressed air energy storage and pumped hydro as potential resource options. Though many studies identified demand response as a potential resource in portfolios, none of the studies appeared to directly consider the role of demand response in increasing the value of solar.

4.7. Designing Candidate Portfolios for Planning Studies

Although the overall framework used by many LSEs in planning studies can capture many drivers of the economic value of solar, ultimately only portfolios created and evaluated within this framework can be selected as a preferred portfolio. A number of LSEs used detailed methods to *evaluate and select* the preferred portfolio from the various candidates, but they did not always use such sophisticated methods to *create* candidate portfolios in the first place. Creating candidate portfolios for further study, Step 2 described earlier in Section 3.2, is critical to determining if a portfolio with solar (and/or other resources) is the least-cost or most attractive portfolio. A similar point was made by Wiser and Bolinger (2006). As such, to the extent that LSEs decide to include solar as a key potential future resource option, they should ensure varied amounts are included in a diversity of candidate portfolios, with different solar configurations and different technologies. Only by including a wide diversity of well-constructed candidate portfolios with solar can the costs and benefits of solar be properly evaluated.

In particular, studies that manually create relatively few portfolios (often less than 10) for further analysis may be ignoring potentially promising portfolios with lower PVRR or risk. The earlier simple example presented in Section 3.6 illustrates the importance of using well-designed portfolios to identify the potential contributions of solar. If portfolios are manually created, it is better to create a wide range of portfolios with various combinations of resources rather than evaluating few portfolios. Roughly half of the planning studies reviewed did not use a capacity expansion model at any point in the study to assist in the creation of candidate portfolios. One of these planning studies, as further illustration of the point, only evaluated high solar penetration in a portfolio that already included other low-carbon (but potentially high-cost) resources. A portfolio with a mix of solar and fossil fuel resources might have been more attractive than this low-carbon portfolio and perhaps even lower cost than the preferred portfolio. Without first creating and then evaluating such a portfolio, however, it is impossible to make the comparison. To address this, LSEs can use capacity-expansion models to create candidate portfolios or guide the creation of candidate portfolios instead of doing it manually.

4.7.1. Commercially available capacity-expansion models used in planning studies

A number of LSEs used commercially available capacity-expansion models to create feasible portfolios (Table 7). These models appear to be useful for automatically screening hundreds or even thousands of combinations of resource options to identify the least-cost option for a given set of deterministic assumptions about the future (i.e., assuming future parameters are known with perfect foresight).

In addition, "bookend" portfolios were then sometimes created using capacity-expansion models by identifying least-cost portfolios for extreme assumptions about the future. For example, rather than manually creating a low-carbon portfolio, a capacity-expansion model was used to identify a least-cost portfolio out of the large number of potential resource combinations under an assumption of very high carbon prices. In some cases, several different least-cost portfolios were created by the LSE by rerunning the model with different future assumptions, and then these candidate portfolios were evaluated in greater detail (Step 3, Section 3.3) and compared (Step 4, Section 3.4).

Though the use of capacity-expansion models to create candidate portfolios is likely superior to manual creation of candidate portfolios, it is important that the capacity expansion models have the capability to recognize the costs and benefits of various resource options. Some commercial capacity-expansion models could be improved by increasing the temporal resolution and better accounting for operational constraints, which affects the energy value of resources and in turn impacts the degree to which the models will find the inclusion of a resource attractive in terms of minimizing PVRR. A previous review of commercial and non-commercial models available to identify the sizing and placement of storage, including thermal storage, made similar recommendations for improving currently available software (Hoffmann et al. 2010). Sioshansi et al. (2012) similarly noted limitations of capacity-expansion models in recognizing the full value of bulk power storage, particularly with respect to the dynamic benefits associated with the provision of AS and ramping over various time frames. Proper representation of the temporal generation profile of solar, the capacity contribution of solar (and changes in that contribution with penetration), and the integration costs of solar in capacity-expansion models are important to valuing solar correctly in these models. Similarly, accurate accounting of the capabilities of thermal storage and natural gas augmentation is important for CSP. A growing body of research examines methods for incorporating operational constraints and high temporal resolution into models that can be used for capacity expansion without greatly increasing the computational complexity (e.g., Müsgens and Neuhoff 2006, Neuhoff et al. 2008, Palmintier and Webster 2011, Staffell and Green 2012). Such improvements may not be in commercially available models used by many LSEs, and are areas for possible future improvement.

An alternative to modifying the capacity-expansion model software is to estimate adjustment factors for various technologies based on an evaluation in more detailed production cost models or other models that account for these factors. Differences in the value of a resource evaluated using a detailed production cost model compared to the value estimated with the simpler capacity-expansion model could be applied as adjustment factors to resources in the capacity- expansion model. PSCo, for example, developed estimates of coal cycling costs with increased wind penetration (amounting to an incremental cost of about $2.2/MWh when increasing wind from 2 to 3 GW) to include in the Strategist capacity-expansion model (Xcel Energy 2011). Without these adjustments, the cost of cycling coal plants would not have been included in the evaluation of wind resources in the capacity-expansion model. Done appropriately, these adjustment factors should always be tailored to address the gaps in an LSE's modeling approach or capacity-expansion model.

A further improvement to the commercial capacity-expansion models would be to account for risk (and uncertainty) directly in creating a portfolio with the lowest expected cost (as opposed to identifying a portfolio that is only least-cost under the particular set of assumed deterministic model inputs). The approach used by some LSEs already emulates this decision-making process through multiple steps. The NPCC approach, however, appeared to be unique in considering uncertainty in the creation of the optimal portfolio from a single capacity-expansion model. In contrast to other models that assume perfect foresight, the NPCC's in-house Regional Portfolio Model identified an expected least-cost portfolio assuming uncertainty in future parameters like fuel and carbon prices. The model was further used to identify the least-cost portfolio for a specified level of acceptable risk exposure. NPCC refers to this process as "risk-constrained, least cost planning." The NPCC model is similar to an advanced stochastic capacity-expansion planning approach described by Jin et al. (2011).

4.7.2. Ranking resource options based on net cost

For LSEs that cannot rely on commercial capacity-expansion models to create feasible portfolios, simple methods to identify which resources are most likely to minimize PVRR can be used to rank potential resources. We derive a method to calculate how including any resource in the least-cost portfolio affects the PVRR, based on finding the "reduced cost" of the resource capacity variable in the optimal solution to the planning problem (see Appendix B for details). The term "reduced cost" is used in linear optimization (Bertsimas and Tsitsiklis 1997), but in this document the resulting metric will be referred to as the "net cost" of a resource for consistency.

Table 7. Capacity-expansion models used by LSEs considering solar LSE/planning entity Capacity-expansion model

LSE/planning entity	Capacity-expansion model
Duke Energy	System Optimizer, Ventyx
El Paso	Strategist, Ventyx
NPCC	Regional Portfolio Model[28]
PacifiCorp	System Optimizer, Ventyx
PNM	Strategist, Ventyx
PSCo	Strategist, Ventyx
TEP	Capacity Expansion,[29] Ventyx
Tri-State	System Optimizer, Ventyx

LSEs differ in what factors are important in the screening of resources based on net cost. For illustration, we focus on a simple form of this formula that only considers the capacity value and energy value assuming perfect foresight. Based on the same concept, the net cost formula can be expanded to include other factors in the avoided cost such as T&D costs, integration costs, or risk-reduction benefits. In the simplest form, this formula is as follows:

Net cost ($/MWh) = Resource Delivered Cost ($/MWh) Avoided Costs ($/MWh)

Where:

- Net cost ($/MWh) is the change in the expected PVRR if a resource were required to be included in the preferred portfolio.
- Resource Delivered Cost ($/MWh) is the cost the LSE would pay to obtain the power from the resource, delivered to the LSE's loads.
- Avoided Costs ($/MWh) is the combination of the capacity value ($/MWh) and the energy value ($/MWh).
- Capacity value ($/MWh) is the avoided cost of capacity based on the capacity contribution of the resource toward a resource adequacy requirement:

$$Capacity\ Value\ (\$/MWh) = \frac{FC_{peak}}{8760\ h/yr}\frac{CC_s}{CF_s}$$

- Where:
  - FC_{peak} is the annualized fixed investment cost of a peaker plant ($/MW-yr).
  - CC_S is the capacity credit of the solar plant (% of name plate).
  - CF_S is the capacity factor of the solar plant (%).
- Energy value ($/MWh) is the avoided production cost based on the coincidence of generation from the resource and wholesale power prices (or system lambdas):

$$Energy\ Value\ (\$/MWh) = \frac{\sum_t p^t E_s^t}{\sum_t E_s^t}$$

- Where:
  - p^t is the hourly wholesale energy price or marginal production cost (a.k.a. the system lambda, $/MWh).
  - E_S^t is the hourly solar production (MWh).

This same concept can be applied to calculate the net cost of a conventional generation option. The primary difference between the calculation for a conventional resource and a solar plant is in the estimation of the energy value (as described in Appendix B).

Ranking resources using this net cost method has the advantage of being defined with respect to the objective of minimizing PVRR. In addition, this approach does not require defining "back- up" capacity adders or other adjustments that are not otherwise part of the broader LSE portfolio creation or evaluation process.

The formula used to estimate the net cost above is similar to the ranking cost formula used in practice by the CA IOU process and LADWP to create candidate portfolios (Figure 9). A similar ranking approach was used to highlight economically attractive resources in the California Renewable Energy Transmission Initiative (RETI) (Black & Veatch 2008) and the Western Renewable Energy Zone Initiative (WREZ) (Mills et al. 2011). The net cost ranking uses somewhat similar logic to the approach advocated by Joskow (2011). This approach to ranking resources based on the net cost closely parallels the process described in Section 3.6, where the net cost of a resource is the difference between its levelized cost and its avoided cost. This approach also mirrors the economic evaluation process that a number of LSEs appear to follow during the evaluation of bids in the resource procurement process (described in the next section). Based on this, the net cost ranking approach seems to be a defensible and understandable way to screen and rank resources during the creation of candidate portfolios, especially when more advanced capacity expansion models are not available to serve that purpose.

4.7.3. Ranking resource options based on other approaches used in practice

In a few cases, LSEs compared resource options based on their levelized costs, along with a number of cost adders that can make certain resources appear less attractive (Figure 10). The adders include transmission costs, integration costs, and capacity-related adjustments. The transmission cost adder simply adds the estimated cost of delivering the resource to the LSE. Transmission cost adders are higher for distant resources with low line utilization relative to resources assumed to be close to the LSE (or with high line utilization).

$$\begin{aligned}&Energy\ Value\ (\$/MWh) = \\ &\quad \Sigma[(Energy\ Value\ in\ Time\ Period) \times (Energy\ Output\ in\ Time\ Period)]\ / \\ &\qquad\qquad Total\ \ Energy\ Output\end{aligned}$$

$$\begin{aligned}&Capacity\ Value\ (\$/MWh) = \\ &\quad (Dependable\ Capacity\ Factor) \times (Baseline\ Capacity\ Value)\ / \\ &\qquad (Project\ Capacity\ Factor\ \times\ 8760/1000)\end{aligned}$$

Figure 9. Capacity value and energy value formulas used by LADWP to rank resources.

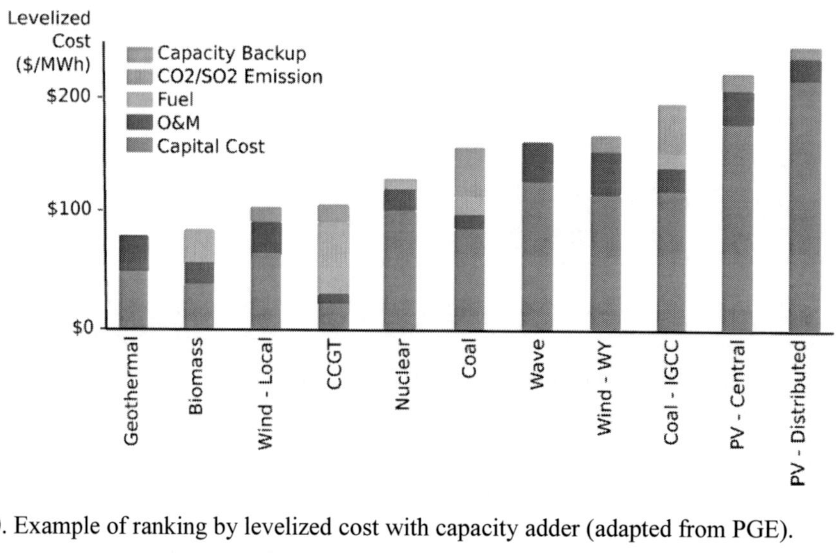

Figure 10. Example of ranking by levelized cost with capacity adder (adapted from PGE).

Integration cost adders are typically based on the uncertainty and variability of variable generation technologies. When included and explicitly stated, the integration cost adders used in resource ranking ranged from $2.5/MWh for APS to $8.25/MWh for NPCC (based on previous estimates of the integration costs for *wind* in that region). One outlier was SRP's very high integration cost adder of $45/MWh applied to the cost of PV. No clear justification was presented for this high integration cost, nor does this high cost appear to be supported by other analysis in the literature.

In a few cases, particularly as LSEs were justifying which resources to include in candidate portfolios, LSEs translated the lower capacity credit of solar relative to its nameplate capacity into a capacity cost adder. PGE based this capacity cost adder on the cost of CTs that would be required to make the capacity contribution of PV and CTs equivalent to the capacity contribution of an energy-equivalent comparator plant (Text Box 2). PSCo used somewhat similar logic in its adjustment of the levelized cost of all resources into an energy- and capacity-equivalent basis (Text Box 3). While the energy-equivalent adjustment approach has been described and employed previously in the broader literature with wind energy, it not as intuitive or as clearly linked to the objectives of the planning studies as the estimate of the net cost of a resource. In contrast, the net cost approach can be clearly linked to the objective of finding portfolios with the lowest PVRR.

TEXT BOX 2. COMPARISON OF CAPACITY COST ADDERS USED BY LSEs TO ADJUST LEVELIZED COST

The basis for the capacity-based adders used by LSEs to screen resources for possible inclusion in candidate portfolios was usually not well described in the planning documents, aside from reference to the lower capacity credit of solar relative to its nameplate capacity. One LSE that did justify its capacity adder to the levelized cost of PV in ranking the resource options was PGE. It adjusted the levelized cost of wind and PV to make the capacity comparable to an energy-equivalent comparator plant (CCGT). The resulting *energy-equivalent* capacity adder for PV with a 5% capacity credit was about $10/MWh. The capacity cost adjustment based on an energy-equivalent comparator plant has similarly been used in a detailed review of the wind integration literature (Gross et al. 2006) and applied to a comparison of wind and a thermal power plant (Söder 2005).

In other planning documents where the basis for the capacity adder was not as clearly justified, the adders ranged from $25/MWh for APS to $47/MWh for TEP, even though the capacity credits for solar in those cases were greater than the 5% capacity credit for PV assumed by PGE. In contrast to the energy-equivalent adder used by PGE, the high-cost examples from APS and TEP appear to be based on the cost of additional capacity resources required to bring solar from its normal capacity credit to a capacity credit of 100%. We refer to this as the *capacity-equivalent* capacity adder to highlight that it makes the contribution to resource adequacy the same as a conventional plant, but the energy generated by the resource is not equivalent to any other particular conventional plant.

To illustrate the differences between an energy-equivalent capacity adder and a capacity-equivalent capacity adder, we walk through a calculation of these adders using assumptions that appear similar to those used by LSEs in particular planning studies.

Energy-Equivalent Capacity Adder

PGE adjusted the levelized cost of each resource option to make the resource comparable to an energy-equivalent comparator plant (CCGT). It started with noting, for example, that a 100-MW CCGT operated at a 92% capacity factor would provide 800 GWh/yr of energy and contribute 100 MW toward the resource adequacy requirement. A 540-MW PV plant with a 17% capacity factor would also provide 800 GWh/yr of energy, but the PV plant would only provide 27 MW of capacity owing to PGE's assumed 5% capacity credit for PV. To have a PV plant that produces 800 GWh/yr of energy also contribute 100 MW toward resource adequacy, an additional 73 MW of CT capacity would need to be added to the portfolio. At a CT cost of $100/kW-yr, the cost of the additional 73 MW of capacity would add $9/MWh to the levelized cost of PV—similar to the capacity adder used by PGE.

Capacity-Equivalent Capacity Adder

In contrast, the capacity cost adder for PV used by TEP appears to be based on adding the cost of 67 MW of additional CTs for each 100 MW of PV with a 33% capacity credit. Assuming the cost of capacity is $100/kW-yr and that TEP would need to add 67 MW of capacity for each 100 MW of PV, the capacity adder would work out to be $45/MWh for PV—similar to the $47/MWh capacity cost adder actually used by TEP.

Although there is disagreement between the LSEs regarding the calculation of capacity adders, the capacity-equivalent capacity adders appear to be higher than can logically be justified since the values do not follow the type of behavior that would be expected based on other detailed analysis of the economic value of solar.

In other studies, LSEs developed capacity cost adders that appear to be much higher than would be expected using an energy-equivalent comparator plant approach. These studies did not clearly specify how they developed the capacity cost adders, but this high cost is not reconcilable with literature on the capacity value of solar at low penetration levels (Text Box 2). These LSEs could develop more representative rankings of eligible resources by shifting to a net cost method (e.g., the current practice for the CA IOU process and LADWP) or at least an energy-equivalent method for comparing different resource options (e.g., the current practice for PGE and PSCo). For clarity and consistency with the objectives of the planning studies, we recommend the use of net cost for ranking resource options.

The risk-reduction benefits of solar were not included in any of the methods used to rank resources when LSEs created candidate portfolios for evaluation. In contrast, one LSE (SRP) applied a risk cost adder to solar and other technologies based on uncertainty about future solar capital costs. This cost adder appears to be unfounded given that capital costs will be relatively well known at the time contracts are signed to build or procure a solar resource. Costs such as natural gas prices or carbon prices in 10–15 years, on the other hand, are comparatively much more uncertain, and power plants must be procured before resolving uncertain future fuel and carbon prices.

TEXT BOX 3. PSCO'S ENERGY- AND CAPACITY-EQUIVALENT LEVELIZED COST APPROACH

PSCo used a commercial capacity-expansion model to create its baseline, least-cost portfolio for a given set of assumptions about the future. It then developed eight alternative renewable portfolios to compare to the least-cost portfolio. In deciding which resources to add to those alternative portfolios, PSCo ranked all potential resources based on adjusting the levelized cost of each resource. The adjustments attempted to account for differences between renewable technologies by putting each of the levelized cost estimates on an energy- and capacity-equivalent basis.

The adjustments to the levelized cost of each resource proceed as follows. First, the resources are made to each produce the same amount of equivalent energy over the year, equal to the annual output of an annual flat block of power with the same nameplate capacity as the resource, by adding "system energy" to the energy produced by the resource. The cost of the system energy is estimated as the energy from a CCGT with a 7,000 MMBTU/MWh heat rate and a specified natural gas price (which works out to be $50/MWh). Then the resources are made to produce the same amount of capacity by adding sufficient new CTs to give the resource a capacity credit equivalent to its nameplate capacity. The total cost of the resource plus the system energy and the additional CT capacity are then divided by the annual energy from the flat block to arrive at the adjusted levelized cost. Table 8 shows the results of this adjustment from the appendix of the PSCo planning study. In this case, the levelized cost of PV is far greater than the levelized cost of wind, but PV has a higher capacity credit per unit of energy than wind. The energy- and capacity-equivalent levelized cost is therefore about equal between wind and PV.[i]

Table 8. Energy- and capacity-equivalent levelized cost example (adapted from PSCo)

Resource	Capacity factor (%)	Capacity credit (%)	Levelized cost ($/MWh)	Energy- and capacity-equivalent levelized cost ($/MWh)	Ranking (lower is better)
CSP with thermal storage (10% ITC)	38	100	223	115	5
PV (30% ITC)	30	55	102	69	3
Wind (no PTC)	45	12.5	76	69	3
CCGT	45	100	81	64	2
CT	10	100	160	60	1

[i] One limitation with this approach is apparent from the comparison of a CT and a CCGT on an energy- and capacity-equivalent basis. In general, a CT has a lower fixed cost than a CCGT, but a CCGT is more efficient. If an LSE needs a new intermediate or baseload generator, the total costs will be lower if the LSE pays a higher fixed cost to buy the CCGT since it will then be able to produce lower-cost energy.

On the other hand, if the new generator will only be needed in rare instances, the total costs will be lower if the LSE pays the lower fixed cost to buy a CT and only uses a small amount of fuel to run the CT infrequently. The PSCo energy- and capacity- equivalent levelized cost, however, assumes that the LSE can get unlimited "system power" with the efficiency of a CCGT and low-cost capacity from a CT. Since the capacity contribution of the CT and CCGT is the same but the fixed cost of the CT is lower (and the CT can rely on unlimited cheap system power), the CT will automatically appear to be a better choice than the CCGT based on this capacity- and energy-equivalent levelized cost approach. This may make sense if the LSE has excess, low-cost "system power," but it is not generally applicable to all areas.
ITC = investment tax credit; PTC = production tax credit

4.8. Economic Evaluation of Bids in Procurement Processes

After identifying candidate portfolios, evaluating those portfolios, and selecting a preferred portfolio, a number of LSEs issued RFPs for generation projects that would either be owned by the LSE or would operate under a long-term contract with the LSE (i.e., a power purchase agreement [PPA] or a tolling agreement). These RFPs almost always provided guidance to bidders regarding the approach the LSE would use to evaluate the attractiveness of each bid based on economic and non-economic (commercial readiness of technology, experience of development team, etc.) factors. This section only covers the methods used to evaluate bids based on economic criteria.

4.8.1. Most LSEs in the study sample appear to rank bids based on net cost

Almost all of the RFPs considered in the survey appeared to evaluate the economic attractiveness of bids by estimating the *net cost* of each bid. In some cases the net cost would be calculated directly using approaches similar to those used in the planning studies. Namely, the net cost would be the difference between the PVRR for the LSE's most recent IRP's preferred portfolio and the PVRR if the resource in question were included in the portfolio. Bids with the lowest net cost would be the most economically attractive (but not necessarily the most attractive overall since many RFPs included non-economic factors too).

In other cases, the LSE estimated the net cost as the respondent's bid cost plus any adders less the benefits of the power from that bid. The adders for solar often included an integration cost adder and in some cases a transmission cost adder. The transmission cost would include any transmission expenditure by the LSE to deliver the solar from its point of delivery to the LSE load area. Any transmission expense for the solar energy to be delivered from the point of grid interconnection to the point of delivery to the LSE was typically included by the RFP respondent in the bid cost. Similarly, the integration cost adder was used only for the cost of the LSE integrating the solar into its system, while any charges by other entities to schedule solar deliveries were included by the respondent in the bid cost. The range of integration cost adders reported in RFPs was within the same range used in portfolio creation and portfolio evaluation reported earlier.

The benefits of the power from the bid typically included an energy value and a capacity value. It was often not clear how the capacity value would be calculated in the evaluation process, although many LSEs described how the capacity credit would be estimated for each bid (or what capacity credit would be assigned to each bid). The energy value would often be estimated as the product of the bidder's generation profile and the wholesale power prices estimated from the LSE's most recent planning study. The temporal resolution of the bidder output profiles requested by the LSEs in the RFPs varied. A 24-hour average daily profile for each month (12 X 24) and a full 8,760-hour time series were commonly used.

For RFPs that sought only RECs or allowed unbundled RECs to compete with bidders that would also deliver power to the LSE, the integration and transmission cost adders of unbundled RECs would be zero, and the benefits would also be zero. The net cost of the unbundled RECs would therefore simply be the bid cost of the RECs, which could be directly compared to the net cost of respondents that provided both RECs and power.

None of the RFPs appeared to use a simplistic economic evaluation method of ranking bids from different resources based only on the lowest bid cost. However, some RFP documents did not specify in detail how the LSE would conduct an economic ranking. Southern California Public Power Authority (SCPPA)—an authority that assists LADWP and other public utilities recently subject to California's 33% RPS by 2020 with procuring renewable resources—only requires a bidder to provide a bid price, maximum and minimum monthly capacity factors, seasonal production shapes, and a description of the project's dispatchability to describe the generating characteristics of its proposed project. SCPPA then indicates that it reserves "the right to make an award to an offer with higher than lowest price offered, or the proposal evidencing the greatest technical ability or other measure, if SCPPA determines that to do so would result in the greatest value to SCPPA and its Member Agencies." No indication is provided by SCPPA as to how it would estimate the value of a bid, suggesting that its evaluation approach lacks detailed quantitative comparisons of the economic value of different bids.

In one case, PSCo, the LSE inputs the characteristics of all bids that pass an initial economic screening into a commercial capacity-expansion model (Strategist) and then uses the model to select the portfolio of bids that minimizes the PVRR. In this case, the merits of each bid are never estimated in isolation; rather the bids are only selected if they, in combination with other bid-in resources, minimize the PVRR. The capacity credit of solar resources used in the capacity- expansion model was based on PSCo's own capacity credit study. As mentioned in Section 4.7.1, this approach to using a capacity-expansion model to evaluate the economic attractiveness of individual bids could be designed, as much as is practical, to account for factors that are not evaluated in the capacity-expansion model (e.g., integration costs, dispatchability of thermal storage, etc.).

In contrast to the widespread consideration of risk in evaluating candidate portfolios in the planning studies, risk was not prevalent in the economic evaluation of bids for most LSE procurement practices. One exception was PSCo. Portfolios of resources bid in response to the PSCo RFP were evaluated under a range of sensitivities, including different natural gas fuel prices, carbon prices, and construction cost escalation rates. Other LSEs may wish to consider ways to embed risk evaluation in ranking of bids.

4.8.2. Lack of detail on economic evaluation can hinder respondents' determination of best technology/configurations

Most LSE procurement practices appear to evaluate the economic merits of individual bids based on the LSE's estimate of the net cost (bid cost less the benefits of the power generated by the resource). This overall framework for ranking bids appears to be a best practice.

That said, there is wide variation in the ways that LSEs appear to calculate the net cost (particularly regarding the estimate of benefits to the LSE of the power generated by a bidder) and little detail on the exact methods used by different LSEs. The lack of detail may be in part due to the additional challenges associated with estimating the avoided cost of variable generation like solar relative to conventional generation, a concern highlighted in a previous survey of utility procurement practices (SEPA 2009). The lack of detail may also be due to the practice of some LSEs working closely with promising bidders to identify the most attractive configuration after initially screening bids at a high level using the RFP process. With larger and more competitive solicitations, however, LSEs may not have the opportunity to work closely with all promising bidders to refine their projects after the initial screening. The lack of detail in the RFPs makes it more difficult for a respondent to determine the most attractive configuration for its project (on a net cost basis). Without detail on how the LSE's hourly marginal production cost is expected to vary throughout the year or which hours of the year drive the need for new capacity, it is difficult to know which PV orientations or tracking technologies are most attractive. The bidder can easily estimate the impact of the choice on its own bid cost but cannot as easily ascertain the impact on the overall net cost (bid cost less benefits to the LSE). In the same manner, it is difficult to know whether using multiple hours of thermal storage with CSP will increase or decrease the net cost of a bid.

Several options might help address these concerns:

(1) The LSE could evaluate the avoided cost of several different solar options and technologies in its planning studies relative to its preferred portfolio. The LSE would then make those estimates available to the market in the RFP or in the LSE's planning studies. RFP respondents could then use those estimates of avoided costs and its own estimates of the impact of different solar configurations and technologies on the bid cost to respond to the RFP with a bid minimizing the net cost.

(2) The LSE could publish information detailing how it will estimate the avoided cost of each bid along with the parameters the LSE will use in the bid evaluation so that a bidder can conduct its own estimate of the expected avoided cost for different solar configurations or technologies. This might include indicating the hourly marginal production costs over future years for the preferred portfolio that the LSE will then use to estimate the energy value component of the avoided cost for each bid. It might also include the estimated cost of capacity for the preferred portfolio and the periods of the year that drive the need for additional capacity.

(3) The LSE could publish only a detailed description of the technical approach that it will use in estimating the avoided cost of individual bids. Each bid respondent would be required to recreate the preferred portfolio of the LSE and estimate the avoided cost of its resource.

This last approach most closely resembles the current situation in many RFPs, although there are cases in which the LSE is ambiguous even regarding the technical details of evaluating the avoided cost of individual bids. An earlier review of utility procurement practices related to solar noted that in this approach the developer is often left to quantify and monetize the benefits of its bid to the utility and then highlight those benefits to the utility in its RFP response (SEPA 2009). Essentially, this situation puts the onus on project developers to understand all of the factors potentially affecting the LSE's valuation of a project over a multi-decade planning horizon and then translate this into an estimate of the avoided cost of potential solar technologies or project configurations that could be bid into the procurement process. Without any clear understanding of how the value would vary across options, developers would be more likely to simply minimize the levelized cost of their bid or submit many similar bids with slightly different configurations, which the LSE would then have to evaluate.

The first two options better inform a bidder during its technology selection and configuration process, with the LSE playing a more direct role in informing the market of what type of generation product would be most valuable. The developer could focus on developing bids that maximize the value to the LSE (i.e., minimize the net cost) rather than focusing on recreating the LSE's estimate of economic value.

As an example of what the first option might look like in practice, the California Public Utilities Commission (CPUC) developed and made publicly available a spreadsheet tool that estimated the net cost of various renewable resource options as part of the CA IOU process (E3's 33% RPS Calculator). In this case, however, the tool was only used for developing portfolios of resources that were evaluated in the CA IOU process (the estimates of the benefits or avoided cost of each option in this spreadsheet tool were not used to evaluate bids during resource procurement). Ideally, an LSE estimate the avoided costs of different options, provide those estimates of avoided costs to bidders, and then use those same values in the economic evaluation of bids during the procurement process.

Finally, this first option (with the LSE providing avoided cost estimates for different resource types, configurations, and locations) may be easier to implement than the second option (with the LSE providing all the detailed parameters needed to calculate the avoided cost of any bid) because the second option may require the LSE to publicly release confidential or sensitive market information.

Conclusion and Recommendations

As renewable technologies mature, recognizing and evaluating their economic value will become increasingly important for justifying their expanded use. We found that many LSEs have a framework to capture and evaluate solar's value, but approaches varied widely: only a few studies appeared to complement the framework with detailed analysis of key factors such as capacity credits, integration costs, and tradeoffs between distributed and utility-scale PV. Factors like the dispatchability benefits of CSP plants with thermal storage appear to be quantified only in terms of a higher capacity credit versus other solar technologies. As the cost of building solar decreases, it will become increasingly important to refine estimates of these factors for all solar technologies, refine study methodologies, and communicate those methodologies to developers and generating equipment manufacturers. In summary we found the following:

- *LSEs should seek, over time, to ensure that solar is included in varied amounts, with different configurations, and for different technologies within a diverse mix of candidate portfolios to ensure a full evaluation of the costs and benefits of solar energy.* Only resources that are fully evaluated within the overall framework can be identified as preferred resources. While some down screening of resource options is warranted to reduce the complexity of the portfolio evaluation process, crude screening can prematurely exclude options that could be attractive with more detailed analysis.

- *Many LSEs can improve their design of candidate portfolios, particularly regarding the methods used to rank potential resource options.* Capacity-expansion models that can represent the costs and benefits of solar technologies should be considered for identifying candidate portfolios, if regulatory and budgetary constraints allow. If that option is not available to an LSE, then the net cost ranking method (as further supported in Appendix B) is a reasonable alternative. The net cost represents an estimate of the change in the PVRR from including a particular resource in the portfolio that would otherwise minimize the expected value of the PVRR. This logical connection to the objective of many planning studies and procurement practices makes the net cost approach attractive for ranking resource options or evaluating the economic merits of RFP bids. A similar net cost ranking approach is used by LADWP and the CA IOU process to select resources that make up candidate portfolios. The net cost method is clearly more defensible than ranking resources based only on levelized costs, and it is potentially easier to connect to the portfolio evaluation process than options that start with levelized costs but then make adjustments such as capacity cost adders. In addition, it appears there is disagreement among the methods used to estimate capacity cost adders: some LSEs are using capacity-equivalent capacity cost adders while others use energy-equivalent capacity cost adders.
- *Planning studies account for the capacity value of solar, but many LSEs can improve their estimates of the capacity credit of different solar technologies at varying penetration levels.* Planning studies accounted for the capacity value of solar by reducing the need to provide capacity through other means, often a CT peaker plant. Most studies accounted for the differential load-carrying capability of PV and CSP without thermal storage relative to CSP plants with thermal storage and/or natural gas augmentation. Only APS, however, appeared to account for changes in the capacity credit of solar with increasing penetration levels. Studies that assume a fixed capacity credit will not recognize one of the main benefits of adding thermal storage or natural gas augmentation to CSP plants—the ability to maintain a high capacity credit even with higher solar penetration levels. In general, few LSEs used detailed analysis of their systems or the broader region to support estimates of the capacity credit of solar. LSEs could improve their planning studies through more attention to capacity credit estimates for different solar technologies, configurations, and penetration levels.
- *Most LSEs have the right approach and tools to evaluate the energy value of solar, but improvements remain possible.* The studies accounted for the energy value of solar by analyzing the ability of solar to reduce variable costs with detailed production cost models. These models can account for both the correlation of high

solar production with times of higher production cost in many regions and the potential for the incremental energy value of solar to decrease with increasing penetration. Evaluating portfolios under a wide range of forecasts regarding uncertain parameters like future fuel and carbon costs demonstrates the degree to which solar resources can reduce exposure to these risks. High temporal resolution (hourly) in the production cost model, inclusion of operating constraints on conventional generation resources (ramp rate limits, start-up costs, minimum generation limits, etc.), and attention to the broader wholesale market outside of the LSE's own resources are important for proper evaluation of the energy value of solar, particularly as the penetration of solar increases. These factors are all represented in more detail in production cost models than in capacity-expansion models. Capacity-expansion models can be improved by increasing the fidelity of the model to represent many of the factors accounted for in more detailed production cost models (while recognizing the need to maintain a reasonable computational complexity). In lieu of changing the capacity-expansion models, LSEs can use detailed analysis with production cost models to develop cost/value adjustment factors that improve the representation of different resource options in capacity-expansion models.

- *Most studies account for integration costs related to solar, but few LSEs have conducted detailed studies to estimate those costs.* Most planning studies accounted for the potential increase in operational integration costs with solar, and many can account for impacts related to hour-to-hour changes in solar production/operational constraints when evaluating portfolios with solar in production cost models. LSEs have conducted few actual studies of operational integration costs or impacts, however, resulting in many LSEs relying on rules of thumb, results from studies in other regions, or results from wind integration studies. LSEs could improve the representation of the benefits of solar in their portfolios by conducting refined integration studies specific to their systems or to systems in their region. Any integration cost estimates from these studies would be most useful if they fill gaps or address limitations in what is already included in the production cost models or capacity-expansion models used by the LSE.
- *Transmission and distribution benefits, or costs, related to solar are not often accounted for in LSE studies.* The difference in avoided costs between utility-scale solar and distributed PV are not well known, but as more studies provide insight into these differences, LSEs should consider incorporating that information into their planning studies.
- *Few LSE planning studies can reflect the full range of potential benefits from the addition of thermal storage and/or natural gas augmentation to CSP plants.* Few studies appeared to include detailed analysis of the potential for CSP with thermal storage to shift solar production to times of most value for a particular scenario, provide ancillary services, or reduce the impact of operational constraints from conventional generation. To capture the full benefits, representation of the dispatchability of CSP with thermal storage (and/or natural gas augmentation) could be improved in both capacity-expansion and production cost models. If it is not practical to model directly the dispatchability of these resources in the LSEs'

planning models, then estimates of the benefits of dispatchability, based on separate side analyses, could adjust the costs/benefits of these resources in the planning models. Identifying the full range of benefits may be increasingly important at higher penetrations of variable renewable energy, particularly solar energy.

- *The level of detail provided in RFPs is not always sufficient for bidders to identify what technology or configurations will be most valuable to LSEs.* While solar developers and manufactures have some ability to tailor solar technologies or configurations to the needs of an individual LSE, the tradeoffs between higher bid cost and higher economic value are not clear in many LSE procurement documents. This lack of clearly defined metrics and weightings in RFPs may hamper efforts to ensure the most economically attractive options are bid into an RFP. It could also lead to respondents providing multiple bids with different technology variations that must be evaluated by the LSE. Increased clarity in the methods used by LSEs to evaluate the net cost of bids may increase the attractiveness of particular RFP responses.

Finally, although this review focused on the valuation of solar in planning and procurement, many of the LSEs are considering other renewable technologies, particularly wind. Many of the lessons learned from this analysis and the recommendations apply to the evaluation of other renewable energy options beyond solar.

List of Planning and Procurement Documents Reviewed

Arizona Public Service

Arizona Public Service (APS). 2012. *2012 Integrated Resource Plan.* www.aps.com/_files/various/ResourceAlt/2012ResourcePlan.pdf

Arizona Public Service (APS). 2011. *AZ Sun Requests for Proposals.* www.aps.com/aps/rfp/default.html

CA IOU Process

Southern California Edison Company, Pacific Gas and Electric Company, San Diego Gas & Electric Company. 2011. *System Resource Plan: Order Instituting Rulemaking to Integrate and Refine Procurement Policies and Consider Long-Term Procurement Plans - Joint IOU Supporting Testimony.* CPUC Rulemaking 10-05-006. www.cpuc.ca.gov/PUC/energy/Procurement/LTPP/LTPP2010/LTPP_System_Plans.htm.

Energy+Environmental Economics (E3). 2011. *33% RPS Calculator* www.cpuc.ca.gov/NR/rdonlyres/89A3151D-F3D8-4073-B35F-5DBF41ADE552/0/RPSCalculator_2007_v2_20120404.xlsm.

California Public Utilities Commission (CPUC). 2010. *Attachment 1: Planning Standards for System Resource Plans – Part II Long-Term Renewable Resource Planning Standards.* R.10-05-006. http://docs.cpuc.ca.gov/efile/RULINGS/118672.pdf.

Southern California Edison. 2011. *2011 Request for Proposals from Eligible Renewable Energy Resource Suppliers for Renewable Products.* http://asset.sce.com/Documents/Shared/2011_SCERFPProcurementProtocolv4.doc.

Pacific Gas and Electric (PG&E). 2011. *2011 RPS RFO Attachment K: PG&E's Description of its RPS Bid Evaluation, Selection Process and Criteria.* www.page.com/includes/docs/word_xls/b2b/wholesaleelectricsuppliersolicitation/RPS2011/Attachment_K_LCBF_06102011.doc.

Duke Energy Carolinas

Duke Energy Carolinas. 2011. *The Duke Energy Carolinas Integrated Resource Plan (Annual Report).* September 1. www.energy.sc.gov/publications/2011DukeEnergyCarolinas IRP_Public.pdf

El Paso Electric

El Paso Electric. 2011. *Request for Proposals for Electric Peaking Power Supply and Load Management Resources.* www.epelectric.com/doing-business-with-epe/el-paso-electric-issues- rfp-for-power-sources-2011

El Paso Electric. 2012. *Integrated Resource Plan of El Paso Electric Company for the Period 2012-2031.* www.epelectric.com/document/integrated-resource-plan-2012-2031-7-16-12

Idaho Power

Idaho Power. 2011. *2011 Integrated Resource Plan.* www.idahopower.com/AboutUs/PlanningForFuture/irp/2011/default.cfm

Imperial Irrigation District

Imperial Irrigation District (IID). 2010. *2010 Integrated Resource Plan.* www.iid.com/Modules/ShowDocument.aspx?documentid=2382

Los Angeles Department of Water and Power

Los Angeles Department of Water and Power (LADWP). 2011. *2011 Integrated Resource Plan.* https://www.ladwp.com/ladwp/faces/ladwp/aboutus/a-power/a-p-integrated resourceplanning/a-p-irp-documents?_adf.ctrl-state=bev0a90il_29&_afrLoop=93913324015000.

Southern California Public Power Authority (SCPPA). 2012. *Request for Proposals for Renewable Energy Projects.* www.scppa.org/Downloads/RFP/RFP_renewablenergy11_30_12.pdf.

Northwest Power and Conservation Council

Northwest Power and Conservation Council. 2010. *Sixth Northwest Conservation and Electric Power Plan.* Council Document 2010-09. www.nwcouncil.org/energy/powerplan/6/default.htm.

Northwest Power and Conservation Council. 2005. *Fifth Northwest Conservation and Electric Power Plan: Appendix L The Portfolio Model.* Council Document 2005-07. www.nwcouncil.org/energy/powerplan/5/default.htm.

NV Energy

NV Energy. 2012. *Nevada Power Company's Triennial Integrated Resource Plan for 2013-2032 and Energy Supply Plan for 2013-2015.* Public Utilities Commission of Nevada. Docket No. 12-06. http://pucweb1.state.nv.us/pdf/aximages/dockets_2010_thru_present/2012-6/17850.pdf.

NV Energy. 2011. *Supplemental Filing on Rehearing continued - Economic Technical Appendix Part 1 of 2. (Volume 3 of 4).* Public Utilities Commission of Nevada. Docket 11-03014. http://pucweb1.state.nv.us/PDF/AxImages/DOCKETS_2010_THRU_PRESENT/2011 - 3/12128.pdf.

NV Energy. 2010. *Spring 2010 Request for Proposals for Renewable Energy and Portfolio Energy Credits.* www.nvenergy.com/company/doingbusiness/rfps/images/2010RFP_Bid_Protocol_Final3-17- 10.pdf.

PacifiCorp

PacifiCorp. 2011. *2011 Integrated Resource Plan.* www.pacificorp.com/es/irp.html

PacifiCorp. 2010. *Request for Proposal: Solar Resources (RFP 2010S).* www.pacificorp.com/content/dam/pacificorp/doc/Suppliers/RFPs/RFP_Solar/RFP2010S_Main-Document-11-30-10.pdf

Portland General Electric

Portland General Electric (PGE). 2009. *Integrated Resource Plan 2009.* www.portlandgeneral.com/our_company/energy_strategy/resource_planning/docs/irp_nov2009.Pdf.

Portland General Electric (PGE). 2012. *Requests for Proposals: Renewable Energy Resources*(Draft). https://portlandgeneralrfp.accionpower.com/_pge_1202/home.asp.

Public Service of Colorado

Public Service of Colorado (PSCo). 2011. *2011 Electric Resource Plan.* CPUC Docket No. 11A- www.xcelenergy.com/About_Us/Rates_&_Regulations/Resource_Plans/PSCo _2011_Electric_R esource_Plan.

Public Service Company of Colorado (PSCo). 2011. *2011 ERP All Source Solicitation: Renewable Resources: Request for Proposals.* www.xcelenergy.com/staticfiles/xe/ Regulatory/Regulatory%20PDFs/PSCo-ERP- 2011/Attachment-3.2-1-Renewable-RFP.pdf.

Public Service of New Mexico

Public Service of New Mexico (PNM). 2011. *Electric Integrated Resource Plan 2011-2030.* www.pnm.com/regulatory/pdf_electricity/irp_2011-2030.pdf

Public Service of New Mexico (PNM). 2011. *Request for Proposals for Renewable Energy Resources or Renewable Energy Certificates (RECs).* www.pnm.com/rfp/ renewablerfp- 2011/docs/rfp_renewablerfp-2011.pdf.

Salt River Project

Salt River Project (SRP). 2010. *Fiscal Year 2011 Resource Plan.* www.srpnet.com/about/ pdfx/ResourcePlanFY2011.pdf.

Coggins, J. 2009. *Resource Technology Options.* SRP RP Workshop. August, 24. www.srpnet.com/about/pdfx/TechnologyOp.pdf.

Tri-State Generation & Transmission:

Tri-State Generation and Transmission Association, Inc., 2010. *Integrated Resource Plan/Electric Resource Plan.* www.tristategt.org/ResourcePlanning/documents/Tri-State_ IRP- ERP_Final.pdf.

Tucson Electric Power

Tucson Electric Power Company (TEP). 2012. *2012 Integrated Resource Plan.* http://ir.uns.com/tep.cfm.

APPENDIX A. SUMMARY OF LSE PLANNING AND PROCUREMENT PRACTICES

Northwest Power Conservation Council

The Northwest Power Conservation Council (NPCC) develops a resource strategy to guide the planning decisions of the Bonneville Power Administration and utilities in the Northwest, a region characterized by significant amounts of energy-constrained hydro resources. The strategy outlines resource types and priorities rather than specifying a particular timing and quantity of power plants to procure. NPCC develops this resource strategy using a regional capacity- expansion model developed by the NPCC. The model develops thousands of portfolios (2,000– 5,000 individual portfolios) that are then each subjected to 750 alternative futures (between 2010–2030) to determine the present value of the total revenue requirement in each of those futures. Each portfolio is characterized by its expected PVRR (the portfolio cost) and the average of the PVRR across the top 10% of most costly futures (the portfolio risk). NPCC then appears to identify the portfolio strategy that is the lowest cost subject to not having higher risk than any of the other portfolios, in what the NPCC calls the "risk-constrained, least-cost plan." This plan represents the least-cost portfolio strategy that can be taken today while limiting exposure to uncertainty in many of the important parameters that impact costs including fuel costs, carbon risk, demand growth, and hydro flows.

The capacity-expansion model does not appear to have peak-hour or annual energy constraints to ensure reliability. Instead, those portfolios that build too little new resources (including energy efficiency or new power plants) are exposed to high market prices in some futures, whereas portfolios that build too many new resources end up being too costly. In this way the overall risk- constrained, least-cost portfolio ends up having significant amounts of new capacity (or energy efficiency) to prevent exposure to bad outcomes relative to what would be built to just satisfy a minimum resource adequacy requirement.

Among other resource options, the NPCC characterizes utility-scale PV (20-MW plants) and CSP with 6–8 hours of thermal storage. Hourly generation profiles from select sites in the Northwest (Ely, NV for CSP) are developed from the NREL System Advisor Model (SAM). The dispatch portion of the capacity-expansion model only uses quarterly average values for both on- peak and off-peak periods to determine production costs for a given portfolio (i.e., the dispatch is characterized by 8 periods per year). The on-peak period is defined as non-Sundays between 7 am and 10 pm in each quarter. NPCC further adds an integration cost to utility-scale PV that is the same as the integration cost added for wind ($8.85/MWh rising to $10.9/MWh by 2024). Although it does not appear that firm capacity is used to develop feasible portfolios in the NPCC capacity-expansion model, the NPCC does indicate that CSP with thermal storage would "impart some firm capacity." Utility-scale PV would only be considered as an energy resource (like wind) due to the poor coincidence of PV and Pacific Northwest loads. No additional detailed modeling is used to evaluate the performance of any of the portfolios. In addition, the NPCC is not responsible for procuring resources identified in the preferred strategy.

Category	Value
Solar technologies considered	Utility-scale PV
	CSP with 6-8 hours thermal storage
Solar capacity credit	Not available
Solar integration cost for resource selection	\$8.85/MWh rising to \$10.24/MWh by 2024 for PV
Solar integration cost for portfolio evaluation	Same
Resolution of production cost model	Two periods per quarter

PacifiCorp

Based on a review of the 2011 IRP and the 2010 Solar RFP, PacifiCorp has one of the more thorough but complex methods for estimating the value and cost of future portfolios of resources and ranking bids received during renewable procurement.

In evaluating future portfolios PacifiCorp begins by using a capacity-expansion model (System Optimizer, Ventyx) to create and select portfolios with the lowest PVRR for a set of assumptions about the future (between 2011–2030). The revenue requirement includes transmission cost, the variable cost of dispatching resources and making market purchases, the fixed cost of maintaining existing units, and the investment cost of any new resources. Each feasible portfolio must have sufficient capacity to meet a minimum planning reserve margin (including the contribution from renewables). The dispatch cost within the model is based on 12 months each represented by a single week. They select optimal portfolios for a given set of assumptions across many different scenarios (more than 60 scenarios). The solar resources included in the analysis include utility-scale PV systems (5-MW plants), distributed rooftop PV, and CSP with and without thermal storage. The CSP without thermal storage has natural gas augmentation with a high heat rate (11,750 Btu/kWh). It is not clear how the reliability contribution of solar is estimated, but the 'Z-method' approximation of the effective load-carrying capability (ELCC) is used for wind (Dragoon and Dvortsov 2006). This approximation does not appear to account for changes in capacity credit as a function of penetration. PacifiCorp also considers many grid-level storage options: batteries, pumped hydro storage, and compressed air energy storage.

These optimal portfolios for a given set of assumptions are then subjected to a Monte-Carlo analysis using a production cost model (PROSYM module within Planning and Risk, Ventyx) to estimate the cost if the portfolio were fixed but the future turned out to be different than assumed when developing the portfolio. The dispatch in the production cost model can account for operational constraints for conventional generation including minimum up and down times, start- up costs, ramp rates, etc. It appears that the data used in the production cost model are based on one week per month for each month of the year. The different portfolios are then compared across several metrics related to cost (average cost), risk (upper tail costs), and reliability (ability to meet demand). The two best-performing portfolios are then evaluated again in 10 different scenarios of carbon cost and natural gas cost (carbon and natural gas are seen as the two largest sources of uncertainty). The final portfolio is then selected from these two.

When PacifiCorp seeks to procure resources that match its preferred portfolio it also uses a thorough evaluation method that builds on its IRP. Bids are ranked based on a net present value metric that is the difference between the value of capacity and energy from the bid and the offsetting costs of the bid. The greater the net value, the higher the bid is ranked. The value of the resources is estimated by comparing the production cost with and without the bid resource included in the most recent IRP preferred portfolio. This production cost savings (the energy value) is then adjusted by any incremental capacity savings and any integration cost from the most recent IRP (no integration cost is specified for solar, but there is a cost for wind). It is not clear how the avoided capacity is estimated for the solar resource.

Category	Value
Solar technologies considered	Utility-scale PV
	Rooftop distributed PV
	CSP with natural gas fired boiler without thermal storage
	CSP with thermal storage
Solar capacity credit	Not available
Solar integration cost for resource selection	Not specified for solar, one is used for wind
Solar integration cost for portfolio evaluation	Same
Resolution of production cost model	Hourly for one week each month, includes operational constraints

Public Service New Mexico

PNM similarly uses a thorough method to estimate the value and cost of future portfolios and to rank bids during renewable procurement.

PNM first uses a capacity-expansion model (Strategist, Ventyx) to create and select portfolios of resources with the lowest present value of the portfolio cost for a given set of assumptions about the future (between 2011–2030). The portfolio cost includes transmission cost, the variable cost of dispatching resources, the fixed cost of maintaining existing units, and the investment cost of any new resources. In contrast to PacifiCorp, the portfolios do not include any off-system sales or imports. Each feasible portfolio must have sufficient capacity to meet a minimum planning reserve margin (including the contribution from renewables). The dispatch cost within the model is based on 12 months each represented by a single week. PNM selects optimal portfolios for a given set of assumptions across many different scenarios (26 scenarios). The solar resources included in the analysis include utility-scale PV systems (40-MW plants), small distributed PV with a lead-acid battery (1-MW plant size), and CSP with three hours of thermal storage and without thermal storage. Forecasts of customer-sited PV are used to reduce the future peak load forecast. The capacity credit for PV and CSP without thermal storage is assumed to be 55% and does not change as a function of penetration. The capacity credit is estimated as the net dependable summer capacity coincident with PNM's system peak load. The capacity credit of the distributed PV with a lead-acid battery is assumed to be 100%. This option was not included in further modeling after a test case conducted using Strategist found a micro-turbine DG plant to be more cost effective than PV with a lead-acid battery.

These optimal portfolios for a given set of assumptions are then subjected to a Monte-Carlo analysis using only the production cost component of the same Strategist model (PROVIEW) to estimate the cost if the portfolio were fixed but the future turned out to be different than assumed when developing the portfolio. The dispatch in this stage does allow for off-system sales or imports (based on wholesale electricity prices generated from sampling historical price distributions). The different portfolios are then compared based on average cost and risk (upper tail cost). PNM selected the least-cost portfolio as its preferred portfolio, noting that in many cases differences in expected costs across portfolio options were greater than differences in the risk across several important uncertainties.

When PNM seeks to procure resources that match its preferred portfolio it also uses an evaluation method that builds on its IRP. Bids are ranked based on a net PVRR metric that is the difference between the costs of the bid and the avoided production costs from the resource. The lower the net revenue requirement, the higher the bid is ranked. The avoided production cost of the resource is estimated by comparing the production cost with and without the resource included in the most recent IRP preferred portfolio. It is not clear whether avoided incremental capacity or integration costs (to account for factors that would not be accounted for in the production cost model) are considered in the ranking.

Category	Value
Solar technologies considered	Utility-scale PV
	Distributed PV with a lead-acid battery
	CSP without thermal storage
	CSP with 3 hours of thermal storage
Solar capacity credit	55% for utility-scale PV and CSP without thermal storage
	100% for small PV with a lead-acid battery
Solar integration cost for resource selection	Not available
Solar integration cost for portfolio evaluation	Not available
Resolution of production cost model	Hourly for one week each month, includes operational constraints

Tri-State Generation & Transmission

Similar to PNM, Tri-State uses a capacity-expansion model (System Optimizer, Ventyx) to create a portfolio of resources with the lowest present value of the portfolio cost for a given set of assumptions about the future (between 2010–2029). Similar to PNM, they select optimal portfolios for a given set of assumptions across many different scenarios (24 scenarios). Using the same scenario-specific assumptions used to create a portfolio, the variable cost of each portfolio is calculated using a production cost model (Planning and Risk, Ventyx) with a higher degree of detail than used in the capacity-expansion model. The production cost model uses hourly data over at least a full year. In some scenarios the production cost model does allow off- system sales or the purchases of power at wholesale electricity prices. The average wholesale electricity price was a scenario-specific assumption. The total PVRR of each portfolio is calculated using the capital cost and fixed O&M cost for existing and new resources from System Optimizer, variable costs from the production cost model, and scenario-specific transmission costs developed by Tri-State's transmission planning group. Tri-State seeks to maintain a 15% planning reserve margin.

The solar resources included in the analysis include PV systems (10-MW plants with no separate consideration of unique benefits of distributed PV) and CSP with three hours of thermal storage.

Tri-State indicates that they use a probabilistic LOLP study to determine the capacity credit of different technologies. For PV, however, they indicate that they rely on the expected capacity of PV during Tri-State's peak load hour. The resulting value is 20% to 57% of the PV nameplate capacity (they do not specify why there is a range of values), but Tri-State does not indicate what value within this range they use, if any, in the capacity expansion model. They do not mention any assumption for the capacity credit of CSP with thermal storage nor do they mention any changes to the capacity credit of solar with penetration. Furthermore, comparison of a scenario with new CCGTs and wind to a scenario with new CCGTs and a combination of wind and CSP with 3 hours of thermal storage indicates that inclusion of solar in the portfolio did not displace the need for capacity from the CCGTs. In both cases nearly 1,200 MW of CCGT capacity is added even though the case with 150 MW of CSP with thermal storage should require less conventional capacity to meet the planning reserve margin. The cause may be the lumpiness of gas-fired plants considered by the capacity-expansion model: the minimum size of gas plants was 290 MW for a 1X1 CCGT. The capital cost of the 290 MW CCGT was 35% more expensive than a larger 588 MW 2X1 CCGT option. The lumpiness potentially prevents the capacity- expansion model from recognizing the capacity contribution of 150 MW of CSP with 3 hours of thermal storage.

Tri-State applies an integration cost to all intermittent resources: the same integration cost is assumed for wind, PV, and CSP with thermal storage in the capacity-expansion model and the production cost model results. In most scenarios the assumed integration cost is $5/MWh. In scenarios with higher proportions of renewables (due to assumed greater RPS levels or carbon policy) the integration cost increases to $7.5/MWh, $8/MWh, or $10/MWh.

Tri-State does not evaluate the sensitivity of the PVRR of each portfolio to changes in assumptions about the future. Tri-State did not select a specific preferred portfolio; instead they used the results of the scenarios to inform the resource procurement plan.

Category	Value
Solar technologies considered	Utility-scale PV
	CSP with 3 hours of thermal storage
Solar capacity credit	Indicate a range of 20%-57% for PV but do not specify the value used
Solar integration cost for resource selection	$5-$10/MWh for PV and CSP with 3 hours of thermal storage depending on the scenario- specific assumptions
Solar integration cost for portfolio evaluation	Same
Resolution of production cost model	Hourly, includes operational constraints

Duke Energy Carolinas

Duke Energy Carolinas includes PV as a potential resource in its IRP (utility-scale PV; it does not appear to separately consider any unique benefits of distributed PV). Duke first applies a basic screening on technology options to determine which should be characterized for inclusion in a commercial capacity-expansion model. The basic screening is done using screening curves (similar to the manner described by Stoft 2002). This approach is useful for dispatchable resources like CTs, CCGTs, coal, nuclear, and biomass renewable resources. Duke Energy, however, also applies the same approach to wind and solar, though it limits the screening curve to the range of plausible capacity factors for each technology. Implicitly, the use of screening curves assumes the resources will be perfectly dispatchable within the available capacity factor of wind and solar. If the screening curve shows the wind and solar is less attractive than other generating options even using this generous assumption then it is clearly not going to be attractive even if the realistic coincidence of generation and load is accounted for. On the other hand, if this screening curve approach does show a lower cost for the wind or solar the result is still uncertain since the real correlation between generation and load is not contained within the screening curve. In this case alternative comparison methods would need to be used such as the net cost approach discussed in the main text. The screening curve approach with wind and solar is therefore subject to the critique from Joskow (2011) that suggests that comparisons on levelized cost are not enough to determine relative economic attractiveness of various options.

After screening technologies, Duke characterizes the selected resource options for use in the System Optimizer (Ventyx) capacity-expansion model. Duke creates a portfolio of resources that satisfies the assumed 17% planning reserve margin with the lowest PVRR for a set of assumptions about the future. The only solar technology that is characterized in the capacity- expansion model is PV. PV is assumed to contribute 50% of its nameplate capacity toward meeting peak load. Duke then generates alternative portfolios by altering the assumptions used in the capacity-expansion model. The sensitivities include changes in fuel prices, load levels, and construction costs. A subset of the portfolios are then chosen for more detailed analysis in a production cost model (the name of the model is not specified nor are many other details about the model). Duke highlights concerns about renewables integration but does not specify an integration cost for PV (or wind) for use in the capacity-expansion or production cost model.

Category	Value
Solar technologies considered	Utility-scale PV
Solar capacity credit	50% for PV
Solar integration cost for resource selection	Not available
Solar integration cost for portfolio evaluation	Not available
Resolution of production cost model	Not available

Tucson Electric Power

Similar to PNM, Tri-State, and Duke, TEP uses a capacity-expansion model (Capacity Expansion [a precursor to System Optimizer], Ventyx) to create a baseline portfolio with reference assumptions and several other portfolios under conditions where assumptions about the future differ from the reference case. For TEP the key assumptions include natural gas prices, wholesale power prices, and load growth. TEP, however, does not present any results or analysis based on these sensitivity scenarios. Instead the baseline portfolio is compared to a small set of manually created alternative portfolios. In the 2012 IRP TEP was particularly focused on the impact of potential coal plant retirement or divestitures. The capacity-expansion model maintains all existing coal plant investments in the baseline portfolio. The alternative, manually created, cases replace the capacity from a coal plant investment with an equivalent amount of capacity from a new combined-cycle gas turbine. Each portfolio is designed to have an equivalent 15% planning reserve margin in all years, including the dependable capacity contribution of solar and wind. No additional assessment of risk or sensitivity to changes in assumptions about the future are presented in the IRP.

Category	Value
Solar technologies considered	Fixed and single-axis PV
	CSP with and without thermal storage
Solar capacity credit	33% for fixed PV
	51% for single-axis tracking PV
	70% for CSP without thermal storage
	87% for CSP with thermal storage
Solar integration cost for resource selection	\$4/MWh for fixed and single-axis tracking PV
	\$2/MWh for CSP without thermal storage
	\$0/MWh for CSP with thermal storage
Solar integration cost for portfolio evaluation	Same
Resolution of production cost model	Not available

The solar resources that are considered as candidate resources to be included in the portfolios included fixed and single-axis tracking PV (both 20-MW plants) and CSP with and without thermal storage. A small amount of distributed PV is included to comply with the DG component of the Arizona renewable energy standard; no additional benefits of distributed PV appear to be identified in the planning study. To better understand the selection of resources in the capacity- expansion model, TEP estimated the relative attractiveness of different resource options. TEP estimated the delivered cost of each resource in \$/MWh terms with the cost for wind and solar adjusted by an integration cost and a capacity cost adder. The integration cost was \$4/MWh for fixed and single-axis tracking PV, \$2/MWh for CSP without thermal storage, and \$0/MWh for CSP with thermal storage. The solar integration costs were based on the previous APS wind study, a PSCo solar integration study, and a Navigant PV integration study for NV Energy. TEP assumes the capacity credit of fixed PV is 33% of the nameplate capacity, single-axis tracking PV is 51%, CSP without thermal storage is 70%, and CSP with thermal storage is 87%. The capacity credits are assumed based on expected generation coincident with peak demand. The resulting capacity cost adder is about

\$47/MWh for fixed-axis PV, \$23/MWh for single-axis PV, \$12/MWh for CSP without thermal storage, and \$4/MWh for CSP with thermal storage (and in the range of \$30–\$35/MWh for wind). It is not clear how TEP estimates the capacity cost adder.

The net PVRR for the baseline and three alternative portfolios is then estimated based on the capital expenditure and the production costs from a production cost model (Planning & Risk, Ventyx). It is not clear what time resolution the model uses in this case. It appears that the same integration costs used in the capacity-expansion model are applied in the production cost model.

El Paso Electric

El Paso Electric also uses Strategist to create portfolios of resources to satisfy its needs for future years (2012–2031). El Paso first uses Strategist to generate an optimal expansion plan based on a baseline set of assumptions. El Paso then generates several additional portfolios based on sensitivity scenarios that change one uncertain assumption at a time (high natural gas price, low natural gas price, higher carbon dioxide prices, lower load forecast, higher load forecast, later retirement years for existing plants). PV is the only solar technology considered in the portfolios (20-MW thin film plants, no separate distinction of the benefits of distributed PV). It is not clear what capacity credit is assigned to PV or how it is estimated. El Paso assumes a constant 15% planning reserve margin. It is also not clear what time resolution is used in the production cost modeling. El Paso indicates that they use typical energy profiles "particularly during EPE's [El Paso Electric's] summer peak months, May through September, to capture the resources' intermittency." El Paso does not mention any adjustments for integration costs for PV.

El Paso selected the portfolio with the least cost in a case with assumed later retirement dates for existing units. El Paso explains that without the later retirement dates El Paso is concerned that too many new units would need to be built at the same time. Later retirement of existing units allows the construction of new generation capacity to be staggered.

El Paso had a recent RFP for peaking resources that allowed renewable resources to participate. El Paso made it clear, however, that any renewable resource would need to be able to be dispatched by El Paso Electric on an hourly basis. Furthermore, renewable projects must also specify the project's minimum guaranteed on-peak generation between 11 am and 4 pm from May through September in order to determine the capacity value. The PPA will contain penalty provisions for not meeting this minimum. As part of the proposal the renewable bidders are required to provide a typical day hourly profile for each month. These stringent conditions may be suitable for CSP with thermal storage or with natural gas augmentation but would likely prevent the participation of CSP without thermal storage or natural gas augmentation and PV without electrical storage. Interestingly, the RFP resulted in a winning bid from a solar project that did not meet the firm capacity requirement along with El Paso's self-bid of four LMS100 CTs.

The methodology used to carry out the economic evaluation of bids is not clear in the RFPs, though El Paso indicates that they consider a resource's relative cost effectiveness in meeting their requirements. Factors included in establishing cost effectiveness include the costs of the resource (capacity costs, energy/fuel costs, fixed and variable O&M costs, start-up costs) and the benefits of the resource (production cost impacts, net capacity contribution).

Category	Value
Solar technologies considered	PV
Solar capacity credit	Not available
Solar integration cost for resource selection	Not available
Solar integration cost for portfolio evaluation	Not available
Resolution of production cost model	Not available

Public Service of Colorado

Like PNM and El Paso, PSCo utilizes the Strategist capacity-expansion model to create a portfolio of resources. Unlike PNM and El Paso, however, PSCo only uses the model to build one least-cost baseline portfolio. The baseline portfolio is built using "starting point" assumptions regarding future fuel costs, investment costs, etc., rather than building many portfolios using Strategist for various scenarios of potential future trajectories. The solar resources included in the portfolio options include PV (25 MW, no separate consideration of the benefits of distributed PV) and CSP with thermal storage. Grid-level batteries (25 MW) are also included as a generic dispatchable resource that can be chosen by the capacity-expansion model. The production cost component uses an hourly generation profile over a week for each month of the year for single-axis tracking PV and CSP with thermal storage. A static hourly generation profile was used to model CSP with thermal storage rather than allowing the thermal storage to be dispatched according to system needs. An integration cost is added to the cost of PV. The most recent publicly available study, from 2009, estimates the integration cost based on the cost related to day-ahead forecast errors ($5.15/MWh). The capacity credit of PV is based on its most recent ELCC study, which estimates a 55% capacity credit for PV. The capacity credit of CSP with thermal storage is assumed to be 100%.

After using the model to build the baseline portfolio, PSCo then manually created eight different renewable alternative plans to compare to the least-cost baseline portfolio. In some cases, the Colorado Public Utilities Commission requires PSCo to adjust the composition of the alternative plans to examine resources that are of interest but may not have been identified as part of the least-cost baseline (e.g., high renewables portfolios). PV is added to a portfolio that already includes much more wind while CSP with thermal storage is added to portfolios that already include PV and wind. Some of the manually created portfolios with wind and PV also included batteries. The addition of the batteries always increased the PVRR relative to the same portfolio without the batteries. The overall reliability of the alternative portfolio is maintained at a constant level by removing CTs from the portfolio based on the capacity credit of the renewables (the CTs are actually lumpy, so CTs end up only being removed when the total amount of renewables in the portfolio is large; any excess capacity is credited at the capital cost of CT).

PSCo then tests the baseline portfolio and the eight alternative portfolios to six sensitivity scenarios where the resources included in the portfolio remain fixed. Two additional sensitivity cases, a high and low sales case, were also examined but the resources in each portfolio were changed to maintain the same level of reliability compared to the median load base case. The same integration cost is added to the production cost results for PV based on the 2009 solar integration study. Only CTs were added or removed from the portfolios in response to changes in projected sales. After reviewing the performance of the different portfolios (using the PROVIEW production cost module within Strategist), PSCo selected the baseline portfolio as the preferred portfolio.

When procuring resources, PSCo places all resource bids that pass an initial economic screening into the Strategist model to develop a least-cost portfolio of bids that minimize the net PVRRs over the planning period (2011–2050). Additional portfolios that represent a range of renewable technologies, PPA lengths, ownership arrangements, etc. are also advanced for further analysis. These portfolios are then subjected to different scenarios regarding natural gas fuel price, carbon price, and construction escalation rates. The preferred portfolio is selected by the utility after reviewing these results. For solar resources the generation profile is based on one of four regional profiles (with or without tracking) and then scaled to the bidder's capacity and energy generation. The solar capacity credit of PV is based on the most recent ELCC study, which varies by region, and whether or not the PV has tracking. The capacity credit does not change as a function of the amount of solar included in the portfolio.

Category	Value
Solar technologies considered	Utility-scale PV
	CSP with thermal storage
Solar capacity credit	55% for PV
	100% for CSP with thermal storage
Solar integration cost for resource selection	\$5.15/MWh for PV
	\$0 for CSP with thermal storage
Solar integration cost for portfolio evaluation	Same
Resolution of production cost model	Hourly for one week each month, includes operational constraints

California IOUs Process - LTPP and LCBF

CPUC uses the Long-Term Procurement Plan (LTPP) process as an umbrella proceeding to comprehensively evaluate and refine various procurement practices and policies related to several other programs including the RPS, energy-efficiency, demand response, resource adequacy, and transmission. When necessary, recommended changes to practices or strategies identified in the LTPP can be implemented via the appropriate proceeding for each particular program. In addition, renewable resource portfolios developed in the LTPP process are used to identify transmission needs in the CAISO transmission planning process. Renewables procurement authorization on a year-to-year basis currently occurs through a different process at the CPUC, but the LTPP is intended eventually to be the place where long-term procurement planning occurs for renewable resources to meet the 33% RPS. As such, while renewables are an integral part of the LTPP process, the ranking of different resource portfolios currently has a limited and somewhat indirect impact on renewable procurement decisions.

In the 2010 LTPP process, the IOUs worked with the CPUC, the California Independent System Operator, and stakeholders to develop and evaluate future resource portfolios that included significant amounts of renewable energy ("the CA IOU process").[30] In contrast to the approaches at other LSEs described in this Appendix that relied on commercial capacity-expansion models to develop resource portfolios given different assumptions, the CA IOU process selected four primary portfolios that were compared across several metrics including the PVRR. Resource selection in the four portfolios was limited to the renewable net short and CTs needed to maintain system reliability. The renewables net short is the difference between the amount of renewables required to meet the 33% by 2020 RPS and the current level of renewables already in place or contracted (renewable resources with signed PPAs, PPAs under review by the Commission, or resources with major permits already granted). Demand-side measures, including energy efficiency and demand response programs, were assumed to be met based on California targets and goals. The selection of the renewables to meet the net short varied across the four portfolios to create bookends that show the impact of different policy objectives (trajectory, cost constrained, environmental impact constrained, and time constrained).

The renewables for the cost-constrained scenario were ranked by the resource-specific economic ranking cost (similar to the RETI/WREZ approach). The potential renewable resources included wind, PV, CSP without thermal storage, geothermal, small hydro, biomass, and biogas. It appears that CSP with thermal storage was characterized, but was not included as an option for the resource ranking or inclusion in the four primary portfolios. The ranking cost is the difference between the resource cost (PPA price, interconnection and transmission delivery cost, and integration cost) and the resource value (capacity value, energy value, avoided T&D costs). The integration cost was assumed to be $7.5/MWh for all wind, PV, and CSP. The capacity value is based on the capacity credit of the resource and the avoided cost of a new CT. The capacity credit in the ranking cost calculation varied by solar technology and location. For PV the capacity credit of utility-scale fixed PV was 51% and utility-scale tracking PV was 65%, but it was assumed to be 0% for small and distributed PV. The capacity credit of CSP without thermal storage ranged from 71% to 87% depending on location, and the capacity credit of CSP with thermal storage was 100%. The energy value is based on the time-varying generation profile and time-varying wholesale energy prices in California over a year. The avoided T&D cost benefit is only applied to distributed resources, including distributed PV, and represents the estimated deferral of T&D network upgrades. The avoided T&D benefit ranged from $4.28/MWh to $26.26/MWh with the most common benefit being $4.84/MWh.

Once the four portfolios were created, sufficient CTs were added to the portfolios in order to meet a planning reserve margin above the expected load in 2020. The capacity credit for each portfolio was estimated based on the California net qualifying capacity (NQC) methodology, which looks at the generation during the peak load period (1–6 pm April–October, 4–9 pm November–March) that is exceeded 70% of the time. The NQC is not equivalent to an ELCC and does not change with penetration.[31] Detailed production cost modeling (PLEXOS, Energy Exemplar) was then used, in part, to estimate the total revenue requirement for each of the portfolios. The production cost modeling included a detailed hourly simulation over a full year of the entire WECC system and included increased operating reserve requirements for each renewable portfolio. Additional CTs were added in cases where in some hours the production cost model found that the resources would not be able to meet all load and operating reserve requirements. The cost of these CTs, CTs added for meeting the planning reserve margin, renewables, existing generation O&M, variable fuel and emissions, and T&D were all included in the estimate of the PVRR. The distribution costs in the case with significant distributed PV was lower than the other cases. Since the production cost model is run for each portfolio, any change with the energy value of different renewables is also captured in the modeling. Further, even though the capacity credit assigned to renewables does not change with penetration, any actual decrease in the true capacity credit would be resolved through the addition of CTs to solve violations in the detailed production cost model runs. No preferred portfolio was selected from the 2010 LTPP process, though the cost-constrained portfolio was found to have the lowest cost. Also in contrast to other LSE studies in the sample, no detailed scenario analysis or stochastic analysis was used to evaluate changes in the revenue requirement if key uncertainties, including natural gas fuel cost or carbon costs, changed from the assumed levels.

When evaluating bids for procurement of renewable resources, the California IOUs rank bids based on a "least-cost, best-fit" methodology (LCBF). The methodology ranks bids based on the costs and benefits of each individual bid. For Pacific Gas and Electric (PG&E) this results in a market value estimate that is the benefits less the costs. For SCE this results in a net levelized cost of renewables premium that is the costs less the benefits. In both cases the benefits include the capacity benefit and energy benefit. The costs include the bid in contract cost, transmission costs, and an integration cost adder. Based on guidance from the CPUC, the integration cost adder is currently assumed to be zero (even though the ranking of resource options in the creation of candidate portfolios described above assumed an integration cost of \$7.5/MWh). The capacity benefit is based on the current estimate of the capacity credit (based on the NQC 70% exceedance methodology outlined in the CPUC resource adequacy program) and estimates of the cost of capacity resources. The energy benefit in the case of SCE is based on the product of the bidding resource's generation profile and hourly wholesale power prices from a production cost model run (PROSYM, Ventyx). The base resource profile for the production cost model run is a recent portfolio from the California long-term planning process. PG&E similarly calculates the energy value based on the bidding resource's generation profile and wholesale market prices, but PG&E uses forward electricity prices rather than prices modeled in a production cost model.

Category	Value
Solar technologies considered	Fixed and tracking PV
	CSP without thermal storage
Solar capacity credit	51% for utility-scale fixed PV
	65% for utility-scale tracking PV
	0% for small scale and distributed PV
	71%-87% for CSP without thermal storag
	100% for CSP with thermal storage
Solar integration cost for resource selection	\$7.5/MWh for PV and CSP without thermal storage
Solar integration cost for portfolio evaluation	Implicitly calculated as increased operating reserve requirement and additional cost of any new CTs required to meet operating reserves
Resolution of production cost model	Hourly over a year, includes operational constraints

Los Angeles Department of Water and Power

Similar to the methodology used by the CA IOUs in the long-term planning process, LADWP adds sufficient renewable resources to its portfolio to meet the 33% RPS by 2020. The renewable resources are selected using a ranking cost for all renewable options. The ranking cost is meant to "measure different renewable resources on a comparable basis" and is calculated in a manner similar to what is used in the CA IOUs process (and similar to the method used in RETI/WREZ). The renewable resources considered include wind, geothermal, biomass, and PV (utility scale or distributed) and were largely based on resources identified within the WREZ process. The ranking cost is the renewable generation cost plus grid integration cost (including transmission and balancing costs) less the capacity value and energy value. The capacity value is based on the capacity credit (called the dependable capacity) and the fixed investment cost of a CT. The capacity credit for PV appears to be set at 27%: there is no suggestion that the capacity credit is based on an ELCC-like method nor that it would change with penetration of solar. The energy value is based on the renewable resource generation profile and time-varying wholesale market prices, though it is not clear what source was used to generate the prices or the generation profiles of the renewable generators.

Aside from the renewable portion of its portfolio, LADWP manually created a small number of different portfolios to test the impact of different decisions regarding divesture of thermal power plants (primarily coal). The makeup of the portfolios was determined based on expert judgment and a constraint that each portfolio must have sufficient dependable capacity (including the contribution of renewables) to meet the projected generation capacity requirement (load plus operating reserves) in each year between 2013 and 2020. A mix of energy efficiency, demand response, and combined cycle natural gas plants was added to each scenario with short-term purchases filling in any remaining need. It is not clear what, if any, cost was assigned to the short-term purchases. The total PVRR for the different portfolios was calculated using a production cost model (Planning and Risk, Ventyx) for three sets of assumptions: a high, reference, and low natural gas price. The production cost model was also used to calculate the CO_2 emissions for each portfolio. A portfolio that had a slightly higher cost but lower CO_2 emissions than other options was selected as the preferred portfolio. The production cost model is chronological, accounts for operational constraints on conventional generation, and has an hourly time resolution, but it is not clear if they run the analysis over 8,760 hours per year or a typical week each month. The model also appears to include some representation of the transmission network, which may lead to additional value in the form of reduced line losses for distributed PV that is sited in the transmission zones that represent major load centers (as long as the production cost model accounts for line losses).

LADWP, like many publicly owned utilities in Southern California, procures renewable energy at least in part through SCPPA. SCPPA's most recent renewables RFP does not provide many details regarding the economic evaluation of proposals. One potentially telling feature, however, is that SCPPA applies a maximum levelized cost cap that differs by technology as part of the initial proposal screening. Wind bids must be less than \$60/MWh, baseload technologies like geothermal must bid less than \$100/MWh, and solar technologies with or without storage must bid less than \$110/MWh. SCPPA requires a bidder to detail its energy availability by describing maximum and minimum monthly capacity factors, seasonal shapes, and dispatchability. SCPPA also requires that all capacity rights associated with the energy are provided to SCPPA. Aside from those requirements SCPPA does not describe how they economically evaluate different proposals or determine which bids are most attractive.

Category	Value
Solar technologies considered	Utility-scale and distributed PV
Solar capacity credit	27%
Solar integration cost for resource selection	Not available
Solar integration cost for portfolio evaluation	Not available
Resolution of production cost model	Hourly with operational constraints

Arizona Public Service[32]

Similar to the California process and LADWP's approach, APS does not use a capacity-expansion model to generate portfolios of resources to meet its needs: portfolios are manually designed. Unlike the California approaches, APS does not use a ranking cost method to determine the relative ranking of renewable resources to include in its candidate portfolio. Instead APS developed four different portfolio options largely based on engineering judgment, stakeholder input, and an assessment of options for maintaining compliance with regulations (including energy efficiency and renewable targets).

APS does include a discussion of the relative "delivered cost" of different resource options that guides the design of the portfolios. The levelized delivered cost of different resources includes the generation cost, emissions cost, and transmission and losses cost. In addition, for renewable resources, the cost includes an integration and firm-up cost. The integration cost for wind ($3.25/MWh) was derived from an APS-specific integration study. The integration cost for PV ($2.5/MWh) was based on assumptions from the Western Renewable Energy Zone Generation and Transmission Model. The firm-up cost, on the other hand, is much larger (>$60/MWh for wind, ~$25/MWh for solar). The firm-up costs are based on the capacity-equivalent capacity adder approach described in this report in Text Box 2. For example, APS assumes a 70% capacity credit for single-axis tracking PV generation such that adding 100 MW of PV requires an additional 30 MW of firming capacity in order for the PV and firming capacity to have a combined load-carrying capability of 100 MW. The firm-up costs are then estimated as the fixed costs associated with the firming capacity (a 30-MW CT) divided by the energy production of the solar project.

In designing each of the four portfolios, the total summer peak dependable capacity is kept constant across all portfolios and is equivalent to the sum of the peak APS demand and reserve requirements. The solar technologies included in the 2012 plan are distributed fixed PV, utility- scale tracking PV, and the existing contract for a CSP plant with six hours of thermal storage (the Solana plant). Other solar technologies were characterized but are not included in any of the portfolios. The line loss reduction benefits of distributed PV are included in the estimate of the costs of portfolios with distributed PV (though APS does not separately summarize these benefits). The base capacity credit of the different solar options is 50% for fixed PV, 70% for single-axis tracking PV and CSP without thermal storage, and 100% for CSP with thermal storage (6 hours) or for solar thermal/gas hybrid plants. Site-specific estimates of capacity credits for single-axis tracking are in the range of 65%–77%, which differs from the generic (non-site specific) estimate of 70%. An ELCC method is used to estimate the capacity credit of solar technologies for each different portfolio of resources (based on the methodology outlined by R.W. Beck 2009). The capacity credit of PV does therefore depend on PV penetration levels in each of the different portfolios (Figure A-1).

The total PVRR for each portfolio is estimated (2012–2027) including the investment costs, fixed costs, variable production costs, and integration costs ($2.5/MWh for PV). The production costs are estimated using a detailed hourly chronological production cost model (PROMOD IV, Ventyx). The PVRR is estimated for each portfolio under various single-variable sensitivity scenarios (high and low natural gas prices, extension of renewable tax credits, high and low energy efficiency costs, and high externality costs). The base case portfolio with the lowest PVRR was selected as the preferred portfolio.

In procuring renewables, APS includes a quantitative analysis of bids that compares the cost of the bidder's proposal to the market cost of comparable conventional generation. The costs include the bid price, any transmission costs to deliver the power to APS, and an integration cost adder of $2.5/MWh for PV ($3.25/MWh for wind). The integration cost is meant to account for the cost imposed on APS of increased resources and regulating reserves due to resource intermittency and forecast uncertainty. The market cost of comparable conventional generation is the avoided capacity and energy cost to APS of producing the incremental electricity taking into account the hourly, seasonal, and long-term supply characteristics of the proposed facility. Bidders are required to submit expected hourly generation profiles (8,760 hr/yr) for the facility based on historical weather years. APS uses production cost modeling of its system to determine the market cost of comparable generation. The capacity credit is estimated using an approximation of the ELCC method where coincidence of generation with top load hours is used to identify the load-carrying capability of a resource. It is not clear how APS evaluates the capacity component of the avoided costs from conventional generation (the capacity value).

Category	Value
Solar technologies considered	Distributed fixed PV Utility-scale single-axis tracking PV CSP without thermal storage CSP with 6 hours of thermal storage Solar thermal gas hybrid plants
Solar capacity credit	Base estimates: 50% for fixed PV 65%-77% for single axis tracking PV 70% for CSP without thermal storage 100% for CSP with thermal storage (6 hours) or for solar thermal/gas hybrid plants These estimates change with penetration of solar in any candidate portfolio
Solar integration cost for resource selection	$2.5/MWh for PV, $0 for other solar technologies
Solar integration cost for portfolio evaluation	Same
Resolution of production cost model	Hourly over a full year with operational constraints

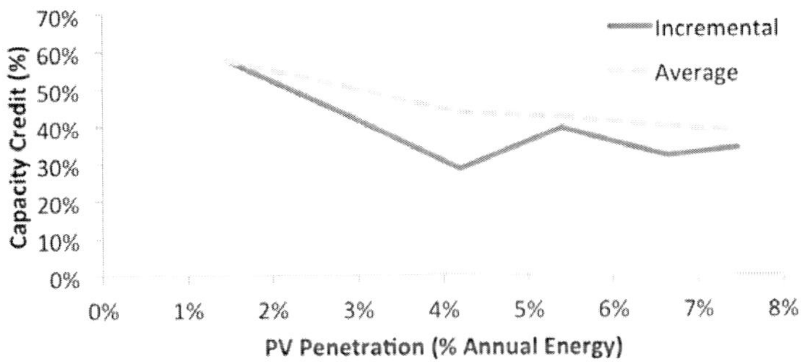

Figure A-1. Estimated PV capacity credit with penetration level for APS Base Plan.

Portland General Electric

Similar to APS, PGE uses engineering judgment to develop different portfolios of resources (15 portfolios). The portfolios are designed to demonstrate extreme bookends where a single resource is chosen to largely fulfill expected future needs (out to 2020) and to then compare those bookends to more diverse portfolios that include combinations of resources. A commercial capacity-expansion model is not used to design the portfolios.

The solar resources that are considered as candidate resources to be included in the portfolios included distributed rooftop PV or ground-mounted utility-scale PV. PGE did not apply separate distinct benefits to distributed PV, but did account for reduced line losses in the production cost modeling. The amount of PV included in each portfolio depended on engineering judgment. The relative attractiveness of different resource options was estimated in order to guide the selection of resources. PGE estimated the delivered cost of each resource in $/MWh terms with the cost for wind and solar adjusted by an integration cost and a capacity cost adder. The integration cost was $6.35/MWh based on the integration cost found for wind at low penetration (the integration cost for wind at the higher penetration levels now experienced by PGE was estimated to be $11.75/MWh in the 2009 wind integration study and revised to $7.96/MWh in the 2011 wind integration study). The capacity cost adder is based on the capital cost of simple cycle CTs needed to make the reliability contribution of solar (and wind) equivalent to the reliability contribution from a CCGT that has the same annual energy output (assuming the CCGT is dispatched as a baseload resource). PGE assumes the capacity credit of PV is 5% of the nameplate capacity (due to PGE being a winter night peaking load). The resulting capacity cost appears to add about $6–$10/MWh to the cost of the PV resources. The same capacity credit is used for wind.

Each portfolio is designed to have an equivalent amount of dependable capacity for various target years, including a 5% capacity contribution of PV and wind. Any remaining capacity necessary to meet the peak hour load and operating reserves after including any demand side measures is filled by simple-cycle CTs and/or on-peak market purchases. On-peak market purchases are limited to meet at most 300 MW of the capacity need in any scenario.

PGE developed a very sophisticated evaluation method to estimate a single metric that can be used to compare the different portfolios across several measures of cost effectiveness, risk and diversity, and reliability. The net PVRR for each portfolio is first estimated under a reference set of assumptions using an hourly production cost model (AURORAxmp, EPIS) and the capital cost of each portfolio (and the integration costs). The portfolios are also then subjected to two types of analysis approaches that evaluate the performance of the portfolios under different assumptions. The first, deterministic analysis varies individual assumptions regarding factors like future fuel prices, carbon prices and carbon price timing, availability of renewable tax incentives, and wholesale market prices. The second, a Monte-Carlo analysis, varies five input variables in a stochastic manner (with no expected correlation between these variables): WECC-wide load, natural gas prices, hydroelectric energy, plant forced outages, and wind production. Reliability metrics are also developed from these stochastic cases based on the amount of energy that PGE is not able to serve with its portfolio and instead relies on market purchases from other WECC resources. The performance of the portfolios in these various cases is then weighted and combined into one ranking score to determine the most attractive portfolio. The weightings are 20% based on reliability and diversity metrics, 30% based on portfolio risk metrics, and 50% based on expected cost.

When procuring renewable resources, PGE considers both price factors (60%) and non-price factors (40%). Both the price factors and non-price factors include information about the generation characteristics of solar. The price score is calculated as the ratio of the total bid cost per MWh to forecast market prices. According to PGE, its price scoring compares the market costs when the energy is delivered to the corresponding hourly projected market price. PGE adjusts the bid price based on whether the bidder includes the cost of providing fixed hourly schedules based on hour-ahead forecasts in order to schedule generation into PGE's system over the transmission network or if the bidder expects PGE to provide this balancing service. If PGE provides the balancing (e.g., the resource is directly connected to the PGE system or is dynamically scheduled to PGE) then PGE adds its expected cost of providing this service to the bid. The integration cost is \$6.35/MWh in \$2009 for solar based on the 2009 IRP integration cost. PGE requests both a monthly 24-hour average generation profile (12 X 24) and a full year of hourly generation data for each solar bid. The 12 X 24 profile represents the monthly energy and peak capacity of the project. PGE does not specify how it accounts for the capacity value of each bid based on the monthly peak capacity.

In addition to the price factors, PGE includes numerous non-price factors in the bid evaluation. Some of the non-price factors include issues related to solar variability and uncertainty. PGE awards the highest non-price score to bids that provide a flat volume of power for all hours. PGE also prefers to know as far in advance as possible the amount of energy to be supplied in any given hour (e.g., week ahead is better than day ahead which is better than hour ahead). PGE does not specify how it assigns a numeric non-price score within these categories, which makes it difficult to know if an inexpensive bid with solar's generation profile and hour-head scheduling would have a higher or lower overall (price and non-price) score than an expensive bid with a flat annual generation profile and week-ahead scheduling.

Category	Value
Solar technologies considered	Distributed and utility-scale PV
Solar capacity credit	5% for PV
Solar integration cost for resource selection	\$6.35/MWh for PV
Solar integration cost for portfolio evaluation	Same
Resolution of production cost model	Hourly over a full year

Idaho Power

Similar to APS and PGE, Idaho Power uses engineering judgment to develop different portfolios of resources (9 portfolios). The portfolios are designed to meet the capacity and energy deficits identified in each future year based on a loads and resource balance along with adding sufficient renewable energy to be able to meet a federal renewable energy standard, if one were to be enacted. A commercial capacity-expansion model is not used to design the portfolios.

The solar resources that were considered as candidate resources for the portfolios included fixed PV (1-MW and larger plant sizes with no consideration of separate benefits of distributed PV) and power towers with 7 hours of thermal storage. The amount of solar included in each portfolio depended on engineering judgment. The relative attractiveness of different resource options was estimated in order to guide the selection of resources. Idaho Power estimated the levelized cost in $/MWh terms (without consideration of integration costs, transmission costs, or any other adjustments) and separately in $/MW of peak hour capacity contribution, or the capacity credit. The capacity credit is estimated based on the capacity factor of each resource during the summer peak period between 3 and 7 pm. The capacity credit of fixed PV was estimated to be 36% of nameplate capacity, and the capacity credit of the power tower with thermal storage was 89%. No integration cost was estimated for the solar resources. Each portfolio is designed to have an equivalent amount of peak-hour capacity for various target years, including the peak hour capacity contribution of solar and wind.

Idaho Power compares the different portfolios based on the net PVRR less any revenues from the sale of excess renewable energy credits. The cost of each portfolio is first estimated under a reference set of assumptions using an hourly production cost model (AURORAxmp, EPIS) and the capital cost of each portfolio. Idaho Power then examines the change in the cost of each portfolio relative to the expected cost by varying the parameter for one uncertainty at a time with both a high case and a low case. The uncertainties include carbon prices, natural gas prices, capital cost, loads, and renewable energy credit prices. Finally Idaho Power used a Monte-Carlo analysis to examine the range and median cost of each portfolio using the same uncertainties as used in the deterministic cases but allowing carbon costs, natural gas prices, and REC prices to be positively correlated. The preferred portfolio had both the lowest expected cost of all portfolios and a relatively low spread of the costs in the Monte-Carlo analysis.

Category	Value
Solar technologies considered	Utility-scale fixed PV and solar power tower with 7 hours of thermal storage
Solar capacity credit	36% for PV
	89% for solar power tower with 7 hours of thermal storage
Solar integration cost for resource selection	Not considered
Solar integration cost for portfolio evaluation	Not considered
Resolution of production cost model	Hourly over a full year

NV Energy

NV Energy is a company formed in the merger of NV Power, which served loads primarily in southern Nevada, and Sierra Pacific, which served loads primarily in northern Nevada. At this time the former Sierra Pacific and NV Power entities file separate IRPs. This review summarizes only the southern Nevada plan covering NV Power's former area. Throughout the document, however, we refer to this as the NV Energy plan. Similar to APS, PGE, and Idaho Power, NV Energy uses engineering judgment to develop different portfolios of resources (four portfolios). NV Energy relies on short-term market purchases to maintain a planning reserve margin when the portfolio resources are insufficient. The short-term market prices include the capacity cost of a CCGT net any short-run profit that CCGT would earn selling its power into the wholesale power market.

The only solar resource considered as a candidate for the portfolios is fixed PV (20-MW plants). NV Energy is currently in a position to meet its RPS requirements with existing and contracted renewable resources. The only portfolio to include PV is a required low-carbon portfolio that includes wind and PV in addition to a common set of conventional generation that is included in all four portfolios (a mix of CTs and combined cycle natural gas turbines). It is not clear how NV Energy determined how much of each generation type to include in the portfolios. The contribution of fixed PV to meeting the planning reserve margin is estimated as 38% of the nameplate capacity. The capacity credit was based on an evaluation of PV generation during peak load periods conducted by Pacific Northwest National Laboratory for NV Energy. No integration cost was explicitly added due to the inclusion of PV in a portfolio, but NV Energy does increase the amount of operating reserve that is held in the production cost modeling based on the nameplate capacity of PV. The amount of increase is based on the results of a detailed PV integration study conducted for NV Energy. NV Energy also adjusted its peak demand load forecast according to their forecast of future customer-sited distributed PV and coincidence of PV output and peak load.

NV Energy compares the different portfolios based on the net PVRR and the net present value of the social cost (which includes the cost of externalities that are not already priced in the PVRR like NOx and particular matter emissions). The cost of each portfolio is first estimated under a reference set of assumptions using an hourly production cost model (PROMOD IV, Ventyx) and the capital cost of each portfolio. NV Energy then examines the change in the cost of each portfolio relative to the expected cost by varying the parameter for one uncertainty at a time. The uncertainties include carbon prices, natural gas and wholesale power prices, loads, and the ability to make off-system sales with NV Energy assets. The preferred portfolio was largely selected based on having the lowest expected revenue requirement and social cost.

In procuring resources NV Energy ranks resources based on relative cost. In the initial screening this includes comparing the proposed project's expected 12 month by 24 hour output profile against NV Energy's avoided cost. After the initial screening, additional detailed economic analysis is carried out. NV Energy does not specify whether its avoided costs include capacity costs nor does it outline what is included in the more detailed economic analysis.

Category	Value
Solar technologies considered	Utility-scale fixed PV
Solar capacity credit	38% for PV
Solar integration cost for resource selection	Not specified
Solar integration cost for portfolio evaluation	Implicitly included through increase in operating reserve requirement in production cost model
Resolution of production cost model	Hourly over a full year with operational constraints

Imperial Irrigation District

IID manually developed five different portfolios of potential future resources to meet a growing gap between existing resources and forecasted peak loads and relatively high reliance on older, inefficient generation. All of the portfolios included a small amount of generic solar resources that would either be met by PV or a solar chimney. IID appears to assign a full 100% capacity credit to the solar resources. IID does not present any detailed cost comparison or ranking methodology to guide the selection of resources for the portfolios.

IID uses an hourly production cost model (PROSYM, Ventyx) to estimate the total revenue requirement of the portfolios under a common set of assumptions about the future. It is not clear how much of each year was simulated in the hourly production cost model. No sensitivity scenario with different assumptions is presented. IID's preferred portfolio is the portfolio with the lowest PVRR.

Category	Value
Solar technologies considered	PV and solar chimney
Solar capacity credit	100%
Solar integration cost for resource selection	Not considered
Solar integration cost for portfolio evaluation	Not considered
Resolution of production cost model	Hourly

Salt River Project

SRP presents a single resource plan without comparing any alternative portfolios or presenting a forecast of the present value of the total revenue requirement of the plan. The plan is designed to meet a gap between existing resources and forecasted future demand while addressing several qualitative objectives. CSP and PV are both considered for the plan, and solar is included in the final plan, but no additional detail on the type of solar technology included in the plan is provided.

When gauging the attractiveness of resources to include in the plan, SRP appears to primarily base the relative merits of each resource option based on its levelized cost of energy plus a category called "integration, delivery, & risk premiums." This includes an integration cost for PV of about \$45/MWh, a \$5/MWh transmission cost for CSP, and a \$40–\$45/MWh capital cost risk for PV and CSP (the wind integration cost is similarly in the range of \$40/MWh).

SRP does not describe the estimated capacity contribution of solar or appear to use a production cost model to estimate the variable costs of the preferred portfolio.

Category	Value
Solar technologies considered	PV and CSP
Solar capacity credit	Not available
Solar integration cost for resource selection	\$45/MWh for PV
Solar integration cost for portfolio evaluation	Not available
Resolution of production cost model	Not available

APPENDIX B. DERIVATION OF NET COST

B.1. Nomenclature

$\mathbb{E}$	Expectation operator
PVRR	Present value of the revenue requirement

Sets:

T	Planning horizon of analysis
Ω	Potential futures

Parameters:

FC	Present value of the fixed cost per unit of capacity (\$/MW-yr)
α_ω	Probability of a future ω
P_s	Cost of involuntary load shedding or hourly cost of buying capacity on the market with short notice, discounted to present value terms (\$/MWh)
MC	Variable cost of producing power, discounted to present value terms (\$/MWh)
ϕ	Availability of a generator in each hour (between 0 and 1)
L	Hourly load
L_p	Peak load and planning reserve margin for resource adequacy
CC	Capacity credit for resource adequacy (between 0 and 1)
CF	Hourly capacity factor of variable generator (between 0 and 1)
E	Hourly energy generation of variable generator ($CF \cdot k$, MWh)

Variables:

g	Hourly generation
l_s	Hourly involuntarily load shedding

s	Slack variable that is positive only when a generator output is less than the full amount available
k	Nameplate capacity of a generation resource (MW)

Dual Variables or Shadow Prices on Constraints:

p	Hourly shadow value of load balance constraint; wholesale power price or system lambda
π	Hourly shadow value of generation capacity limit for each generator
μ_k	Shadow value of resource adequacy constraint

Results:

$\bar{c}$	Reduced cost of a variable ($/MW-yr)
$\bar{C}$	Reduced cost of a variable ($/MWh)

The LSE planner problem is as follows: given a set of uncertain futures, determine the portfolio of resources that leads to the lowest expected present value of the revenue requirement. The solution to this problem is called the preferred portfolio.

B.2. Planning Problem with No Adequacy Requirement

In the most simple case, the planner must determine the portfolio that leads to the lowest revenue requirement without consideration of any adequacy requirement. If insufficient generation is available at any point in time, load will be involuntarily shed or the LSE will need to procure energy from the market on short notice. In either case the cost of this is assumed to be very high. This will tend to push the LSE toward procuring sufficient capacity in the preferred portfolio in order to minimize these high costs.

The planning problem can be mathematically defined as a linear program in standard form as follows:

Objective function:

$$\min \quad \mathbb{E}\,[PVRR]$$

$$= \min \quad FC_v k_v + \sum_i FC_i k_i + \sum_\omega \alpha_\omega \sum_t P_s l_s^{\omega,t} + \sum_i MC_i^\omega g_i^{\omega,t}$$

Load Balance,

$\forall t \in T, \omega \in \Omega$:

$$[\text{dual: } \alpha_\omega p^{\omega,t}] \quad CF_v^{\omega,t} k_v + l_s^{\omega,t} + \sum_i g_i^{\omega,t} = L^{\omega,t}$$

Generator Capacity,

$\forall t \in T, \omega \in \Omega, i \in I:$

$$[\text{dual: } \alpha_\omega \pi_i^{\omega,t}] \qquad -\phi_i^{\omega,t} k_i \qquad + \qquad g_i^{\omega,t} + \; s_i^{\omega,t} \quad = 0$$

Non-negativity:

$$k_v, \qquad k_i, \qquad l_s^{\omega,t}, \qquad g_i^{\omega,t}, \quad s_i^{\omega,t} \quad \geq 0 \tag{1}$$

This same problem can be restated in words as:

Objective Function	Minimize the sum of the annualized fixed cost of capacity and the expected dispatch cost (including any high costs of involuntary load shedding or procuring capacity from the market on short notice)
Load Balance Constraint	In every hour and in every future the generation from the variable generation and the dis- patchable generation must be in balance with the load. If they are not in balance then load must be involuntarily shed or power must be bought on short notice from the market.
Generation Capacity Constraint:	The amount of generation from a generator in any hour must be less than or equal to the amount of available generation capacity in that hour. If it is less than the amount of available generation then the slack variable must be positive.

B.2.1. Solution: Least-cost portfolio without adequacy requirement

Since this problem can be stated as a linear program, there are certain features of the solution to this problem (e.g. the least cost portfolio) that must be true (Bertsimas and Tsitsiklis 1997).

First, every variable will have what is called a "reduced cost". This is defined as the change in the objective value function (i.e., the expected present value of the revenue requirement) for a small increase or decrease in the variable away from the optimal value. The "reduced cost" of a generator that is not included in the preferred LSE portfolio, for example, represents how much the expected PVRR would increase if a small amount of capacity of that generator were forced into to the portfolio and a new least-cost portfolio were found (given the forced-in high cost generation).

This is the same as the definition of the "net cost" used in the main document. When the problem is defined in the standard form for linear programming (Eq. 2), the "reduced cost" for a variable *j* is given by Eq. 3.

$$\begin{aligned} \min \quad & \mathbf{c}'\mathbf{x} \\ \text{s.t.} \quad & \mathbf{Ax} = \mathbf{b} \\ & \mathbf{x} \geq \mathbf{0} \end{aligned} \tag{2}$$

$$\bar{c}_j = c_j - \mathbf{v}'\mathbf{A}_\mathbf{j} \tag{3}$$

In words, Eq. 3 states that the reduced cost of any variable is defined as the cost coefficient of that variable in the objective function less the product of the dual value of each constraint (**v**) and the constraint coefficient for that variable ($\mathbf{A}_j$).

The reduced cost formula can be used to estimate how much the expected PVRR will increase if a variable generator is not part of the preferred portfolio, but the variable generation is added to the portfolio and a new least cost portfolio is found with the addition of that variable generator. The cost coefficient of the variable generation capacity variable in the objective function is the annualized fixed cost of the variable generator (FC_v). Since the only constraint that includes the capacity of the generator is the "Load Balance" constraint, the product of the dual variables of that constraint ($\alpha_\omega p^{\omega,t}$) and the constraint coefficient ($CF_v^{\omega,t}$) is simply the sum of $\alpha_\omega p^{\omega,t}\ CF_v^{\omega,t}$ over all hours and all futures. Since $p^{\omega,t}$ is the dual of the load balance constraint, $p^{\omega,t}$ can be interpreted as the hourly wholesale price for power for a particular future. The product $p^{\omega,t}CF_v^{\omega,t}$ can be described simply as the coincidence of wholesale prices and variable generation. The general formula for the reduced cost, Eq. 3, can then be used to find the reduced cost of the variable generation capacity included in the portfolio (assuming there is no adequacy requirement for the LSE), Eq. 4.

$$\bar{c}_v = FC_v - \sum_{\omega} \alpha_\omega \sum_{t} p^{\omega,t} CF_v^{\omega,t} \tag{4}$$

This result can be interpreted as follows. The amount that the LSE's expected present value of the revenue requirement will increase when variable generation is added to the preferred portfolio is the difference between the annualized fixed cost of the variable generator and the weighted average over all possible futures of the coincidence of wholesale power prices and the output of variable generation.

In some cases it can me more intuitive to understand the annualized fixed cost in terms of the levelized cost per unit of energy (LCOE) instead of the annualized cost per unit of capacity ($/MW-yr). The LCOE is simply the annualized fixed cost divided by the annual generation of a 1 MW variable generation plant ($CF_v 8760h/yr$).

Dividing both sides of Eq. 4 by the annual generation of a 1 MW variable generation plant then leads to the following modified results in the more intuitive $/MWh form in Eq. 5

$$\bar{C}_v = LCOE_v - \sum_{\omega} \alpha_\omega \frac{\sum_t p^{\omega,t} E_v^{\omega,t}}{\sum_t E_v^{\omega,t}} \tag{5}$$

The result is somewhat different if the LSE uses a resource adequacy requirement in determining their preferred portfolio.

B.3. Planning Problem with a Resource Adequacy Requirement

Often LSEs require that any feasible portfolio must satisfy a resource adequacy constraint—the sum of the capacity contribution from resources any portfolio must equal the peak load plus a planning reserve margin—when identifying the preferred portfolio. This results in a slightly modified planning problem as shown below.

Objective function:

$$\min \quad \mathbb{E}[PVRR]$$

$$= \min \quad FC_v k_v + \sum_i FC_i k_i + \sum_\omega \alpha_\omega \sum_t P_s l_s^{\omega,t} + \sum_i MC_i^\omega g_i^{\omega,t}$$

Load Balance,

$\forall t \in T, \omega \in \Omega:$

$$[\text{dual: } \alpha_\omega p^{\omega,t}] \quad CF_v^{\omega,t} k_v + l_s^{\omega,t} + \sum_i g_i^{\omega,t} = L^{\omega,t}$$

Generator Capacity,

$\forall t \in T, \omega \in \Omega, i \in I:$

$$[\text{dual: } \alpha_\omega \pi_i^{\omega,t}] \quad -\phi_i^{\omega,t} k_i + g_i^{\omega,t} + s_i^{\omega,t} = 0$$

Adequacy Requirement,

$\forall t \in T, \omega \in \Omega:$

$$[\text{dual: } \mu_k] \quad CC_v k_v + \sum_i CC_i k_i = L_p$$

Non-negativity:

$$k_v, \quad k_i, \quad l_s^{\omega,t}, \quad g_i^{\omega,t}, \quad s_i^{\omega,t} \geq 0 \qquad (6)$$

The resource adequacy constraint can be restated in words as:

Resource adequacy constraint	The capacity contribution from the variable generators and the conventional generation (de- fined as the capacity credit times the nameplate capacity) must equal the peak load plus planning reserve margin target.

B.3.1. Solution: Least-cost portfolio with an adequacy requirement

The addition of a new resource adequacy constraint changes the formula used to estimate the reduced cost of any variable, including the variable generator, since the new resource adequacy constraint adds both a new dual variable (μk) and a new constraint coefficient for the variable generator (CCv). The change in the expected PVRR with the addition of variable generation is then given by the revised reduced cost, Eq. 7. The change in the expected PVRR must now also account for the capacity credit assigned to the variable generator in the resource adequacy assessment.

$$\bar{c}_v = FC_v - \left(\mu_k CC_v + \sum_{\omega} \alpha_\omega \sum_t p^{\omega,t} CF_v^{\omega,t}\right) \quad (7)$$

In order to interpret this new formula for estimating the reduced cost we must now try to understand the shadow value of the resource adequacy constraint or the dual variable μ_k. First, if the resource adequacy constraint is not bind- ing, meaning that the resource adequacy constraint is satisfied automatically, then the dual variable of the constraint, μ_k, is zero and the reduced cost of the variable generator remains the same as before, without the resource adequacy constraint. On the other hand, when then constraint is binding, the dual variable μ_k will be non-zero and the reduced cost of the variable generator will be different than what would be calculated without consideration of the resource adequacy constraint.

In the case that the resource adequacy constraint is binding, we now turn to the estimation of the numerical value of the dual variable μ_k. Another helpful point in this regard is that for any variable included in the least-cost portfolio, the "reduced cost" of that variable must be equal to 0: if it were negative then adding more of that variable wound decrease the expected PVRR and hence the original portfolio wouldn't have been the least-cost portfolio. Conversely, if the reduced cost were positive then the addition of that variable to the portfolio causes the expected PVRR to increase, no longer making it the least-cost portfolio (since the expected PVRR has increased).This point can be used to better understand the value of μ_k for any generator included in the least cost portfolio.

Using the definition of the reduced cost in Eq. 3, we can calculate the reduced cost of any conventional generation capacity as:

$$\bar{c}_i = FC_i - \left(\mu_k CC_i - \sum_{\omega} \alpha_\omega \sum_t \pi_i^{\omega,t} \phi_i^{\omega,t}\right) \quad (8)$$

For any conventional generation capacity that is included in the preferred portfolio, the reduced cost of that generation capacity $\bar{c}_i$ will be equal to zero. In that case, for any generation capacity included in the preferred portfolio the following will hold:

$$\mu_k CC_i = FC_i + \sum_{\omega} \alpha_\omega \sum_t \pi_i^{\omega,t} \phi_i^{\omega,t} \quad (9)$$

The dual variable for the generator capacity constraint, $\alpha_\omega \pi_i^{\omega,t}$ indicates the degree to which the expected PVRR would change if more capacity of generator i were available in a particular future, ω, at a particular point in time, t. In hours where the generator is dispatched below its available capacity in that hour (such that $g_i^{\omega,t} < \phi_i^{\omega,t} k_i$) then the slack variable indicating how far below the available capacity the generator is dispatched, $s_i^{\omega,t}$, is positive and its reduced cost, $c_{s_i^{\omega,t}}$, must then be zero. Using Eq. 3, the reduced cost of the generator slack variable is:

$$\bar{c}_{s_i^{\omega,t}} = 0 - \alpha_\omega \pi_i^{\omega,t} \quad (10)$$

In those hours when the generator i is dispatched below its capacity, Eq. 10 then implies that $\pi_i^{\omega,t}$ must be equal to zero.

Similarly in any hour when the generator is used to generate power (such that $g_i^{\omega,t} > 0$), the reduced cost of the generator output must be zero. Using Eq. 3, the reduced cost of the generator output is:

$$\bar{c}_{g_i^{\omega,t}} = \alpha_\omega MC_i^\omega - \alpha_\omega p_i^{\omega,t} - \alpha_\omega \pi_i^{\omega,t} \tag{11}$$

In hours when the generation is positive, the reduced cost is zero and the the following must hold.

$$\pi_i^{\omega,t} = MC_i^\omega - p_i^{\omega,t} \tag{12}$$

During the periods when the slack variable is positive and the generator output is zero (indicating the generator is being used but the capacity constraint is not binding) both Eq. 10 and Eq. 12 must hold indicating that the wholesale power price in those hours, $p_i^{\omega,t}$, must equal the variable cost of the generator, MC_i^ω.

From the evaluation of the reduced cost of the generator slack variable s_i and the generator output, g_i, for a particular generator it is clear that time period can be partitioned into two mutually exclusive periods: (1) period T_0 where s_i is positive the such that the generator capacity constraint not binding and $\pi_i^{\omega,t}$ is equal to zero or (2) period T_k where the generator capacity constraint is binding and $\pi_i^{\omega,t}$ is the difference between the variable cost of the generator and the wholesale power price, $MC_i^\omega - p_i^{\omega,t}$. This information can be used to simplify the relationship describing the dual variable of the resource adequacy constraint for a generator that is included in the preferred portfolio, shown earlier in Eq. 9:

$$\mu_k CC_i = FC_i + \sum_\omega \alpha_\omega \left(\sum_{t\in T_0} 0 \cdot \phi_i^{\omega,t} + \sum_{t\in T_k} \left(MC_i^\omega - p_i^{\omega,t}\right) \phi_i^{\omega,t} \right) \tag{13}$$

Simplifying this relationship leads to:

$$\mu_k CC_i = FC_i - \sum_\omega \alpha_\omega \sum_{t\in T_k} \left(p_i^{\omega,t} - MC_i^\omega\right) \phi_i^{\omega,t} \tag{14}$$

This relationship indicates that the dual of the capacity constraint, μ_k , largely depends on the annualized fixed cost of any generation capacity included in the preferred portfolio less the amount that the wholesale power price exceeds the variable cost of that generation.

In a very simplified case where the preferred portfolio includes new peaking generation that is never fully dis- patched to its full capacity, k_{peak} , and the new peaking generation has a capacity credit equal to its nameplate capacity the dual of the resource adequacy constraint would be further simplified to:

$$\mu_k = FC_{peak} \tag{15}$$

Since the new peaking generation is never dispatched to its full capacity, Eq. 10 indicates that π_{peak} , the dual of the peaker plant capacity constraint, is always equal to zero. Since the reduced cost of any variable is greater than or equal to zero (otherwise the preferred portfolio wouldn't be the least cost portfolio) then Eq. 11 also indicates that the wholesale prices must always be less than or equal to the variable cost of the peaker plant.

The change in the expected PVRR with the addition of variable generation in the case that such a peaker plant is included in the preferred portfolio can then be found using Eq. 7 and Eq. 15:

$$\bar{c}_v = FC_v - \left(\overbrace{FC_{peak} CC_v}^{\text{capacity value}} + \overbrace{\sum_{\omega} \alpha_{\omega} \sum_{t} p^{\omega,t} CF_v^{\omega,t}}^{\text{energy value}} \right) \tag{16}$$

This result can be interpreted as follows. The amount that the LSE's expected present value of the revenue require- ment will increase when variable generation is added to the preferred portfolio is the difference between the annualized fixed cost of the variable generator and the sum of the capacity value and the energy value of the variable generator. The capacity value is the product of the annualized fixed cost of a peaker plant included in the preferred portfolio and the capacity credit of the variable generator. The energy value is the weighted average over all possible futures of the coincidence of wholesale power prices and the output of variable generation. The wholesale prices, furthermore, never exceed the variable cost of the peaker plant.

Dividing both sides of Eq. 16 by the annual generation of a 1 MW variable generation plant then leads to the following modified results in the more intuitive \$/MWh form in Eq. 17

$$\bar{C}_v = LCOE_v - \left(\overbrace{\frac{FC_{peak}}{8760\text{h/yr}} \frac{CC_v}{CF_v}}^{\text{capacity value}} + \overbrace{\sum_{\omega} \alpha_{\omega} \frac{\sum_t p^{\omega,t} E_v^{\omega,t}}{\sum_t E_v^{\omega,t}}}^{\text{energy value}} \right) \tag{17}$$

In a more general case where a peaker plant with such stringent requirements (full capacity credit, never fully dispatched) is not included in the preferred portfolio, then the amount that the LSE's expected present value of the revenue requirement will increase when variable generation is added to the preferred portfolio is given by Eq. 18, based on Eq. 7 and Eq. 14. In the more general case, the wholesale power prices may increase above the variable cost of the generator i when the generator output is equal to its capacity.

$$\bar{c}_v = FC_v - \left(\overbrace{\left(FC_i - \sum_{\omega} \alpha_{\omega} \sum_{t \in T_k} \left(p_i^{\omega,t} - MC_i^{\omega} \right) \phi_i^{\omega,t} \right) \frac{CC_v}{CC_i}}^{\text{capacity value}} + \overbrace{\sum_{\omega} \alpha_{\omega} \sum_{t} p^{\omega,t} CF_v^{\omega,t}}^{\text{energy value}} \right) \tag{18}$$

B.3.2. Solution: Conventional generation resource

For any conventional generation resource included in the preferred portfolio, the reduced cost of the generation capacity is zero. The reduced cost of any conventional generation resource that is not included in the preferred portfolio, generator *n*, can be calculated using the same approach as described for the variable generator.

Earlier we started with the equation for the reduced cost, Eq. 3, and then came up with the reduced cost for conventional generation capacity, Eq. 8. For generator *n* we recreate that equation again here for clarity.

$$\bar{c}_n = FC_n - \left(\mu_k CC_n - \sum_{\omega} \alpha_{\omega} \sum_{t} \pi_n^{\omega,t} \phi_n^{\omega,t} \right) \tag{19}$$

For any conventional generator *i* that is included in the preferred portfolio we can find the value of μ_k, the dual of the resource adequacy constraint, using Eq. 14. In the very simplified case where a peaker plant is included in the preferred portfolio, the dual value of the resource adequacy constraint is reduced to FC_{peak} according to Eq. 15.

Similar to the earlier discussion the time period can be partitioned into a period where the slack variable for the generator *n* would be positive, and the dual variable of the generator *n* capacity constraint, $\pi_n^{\omega,t}$, would be zero. In the rest of the year when the generator output is positive and its output is equal to the nameplate capacity limit of generator *n* is binding the dual value of the generator *n* capacity constraint becomes:

$$\pi_n^{\omega,t} = MC_n^{\omega} - p_n^{\omega,t} \tag{20}$$

The reduced cost of the conventional generator that is not included in the preferred portfolio can then be simplified to the following relationship:

$$\bar{c}_n = FC_n - \left(\overbrace{FC_{peak} CC_n}^{\text{capacity value}} + \overbrace{\sum_{\omega} \alpha_{\omega} \sum_{t \in T_n} \left(p^{\omega,t} - MC_n^{\omega} \right) \phi_n^{\omega,t}}^{\text{energy value}} \right) \tag{21}$$

This is essentially the same method for calculating the reduced cost for the variable generator in Eq. 16. The main difference is that the energy value term must now account for the difference between the wholesale price and the variable production cost of the conventional generation source. Further the availability of the conventional generator must also be accounted for.

The reduced cost of the generator per unit of energy produced, $\bar{C}_n$, can also be specified for comparison to the reduced cost per unit of energy from the variable generator.

$$\bar{C}_n = \frac{FC_n}{CF_n 8760\text{h/yr}} - \left(\overbrace{\frac{FC_{peak} \; CC_n}{8760\text{h/yr} \; CF_n}}^{\text{capacity value}} + \overbrace{\frac{1}{CF_n 8760\text{h/yr}} \sum_{\omega} \alpha_{\omega} \sum_{t \in T_n} \left(p^{\omega,t} - MC_n^{\omega} \right) \phi_n^{\omega,t}}^{\text{energy value}} \right) \tag{22}$$

REFERENCES

Arvizu, D., Balaya, P., Cabeza, L., Hollands, T., Jäger-Waldau, A., Kondo, M., Konseibo, C., et al. (2011). "Direct Solar Energy." In *IPCC Special Report on Renewable Energy Sources and Climate Change Mitigation*, ed. O. Edenhofer, R. Pichs-Madruga, Y. Sokona, K. Seyboth, P. Matschoss, S. Kadner, T. Zwickel, et al. Cambridge, UK and New York, NY, USA: Cambridge University Press.

Barbose, G., Darghouth, N. & Wiser, R. (2012). *Tracking the Sun V: An Historical Summary of the Installed Cost of Photovoltaics in the United States from 1998 to 2011*. LBNL-5919E. Berkeley, CA: Lawrence Berkeley National Laboratory. http://emp.lbl.gov/sites/all/files/LBNL-5919e- report.pdf.

Barbose, G., Wiser, R., Phadke, A. & Goldman, v. (2008). "Managing Carbon Regulatory Risk in Utility Resource Planning: Current Practices in the Western United States." *Energy Policy*, *36* (9), (September), 3300–3311. doi:10.1016/j.enpol.2008.04.023.

Bazilian, M. & Roques, F. (2008). *Analytical Methods for Energy Diversity and Security: Portfolio Optimization in the Energy Sector: A Tribute to the Work of Dr. Shimon Awerbuch*. Oxford: Elsevier Science Limited.

Beck, R. W. (2009). *Distributed Renewable Energy Operating Impacts and Valuation Study*. Arizona Public Service. http://www.solarfuturearizona.com/.

Bertsimas, D. & Tsitsiklis, J. N. (1997). *Introduction to Linear Optimization*. Belmont, MA: Athena Scientific.

Black & Veatch. (2008). *Renewable Energy Transmission Initiative: Phase 1A, Final Report*. Sacramento, CA: California Energy Commission. www.energy.ca.gov/2008publications/RETI-1000-2008-002/RETI-1000-2008-002-F.PDF.

Bokenkamp, K., LaFlash. H., Singh, V. & Bachrach Wang, D. (2005). "Hedging Carbon Risk: Protecting Customers and Shareholders from the Financial Risk Associated with Carbon Dioxide Emissions." *The Electricity Journal*, *18* (6) (July), 11–24. doi:10.1016/j.tej.2005.05.007.

Bolinger, M. A. & Wiser, R. (2009). *The Value of Renewable Energy as a Hedge Against Fuel Price Risk: Analytic Contributions from Economic and Finance Theory*. LBNL-2022E. Berkeley, CA: Lawrence Berkeley National Laboratory. http://escholarship.org/uc/item/65g8f2t4.pdf.

Borenstein, S. (2008). *The Market Value and Cost of Solar Photovoltaic Electricity Production*. Berkeley, CA: UC Energy Institute. www.ucei.berkeley.edu/PDF/csemwp176.pdf.

Brinkman, G., Denholm, P., Drury, E., Ela, E., Mai, T., Margolis, R. & Mowers, M. (2012). *Grid Modeling for the SunShot Vision Study*. NREL/TP-6A20-53310. Golden, CO: National Renewable Energy Laboratory. http://www.nrel.gov/docs/fy12osti/53310.pdf.

Burgess, E., Estrada, G. & Mayes, K. (2012). *Transmission and Renewable Energy Planning In California: Opportunities for Regional Stakeholder Engagement*. Montpelier, VT: Regulatory Assistance Project. www.westgov.org/wieb/wrez/11-28-2012WREZca.pdf.

Chu, S. & Majumdar, A. (2012). "Opportunities and Challenges for a Sustainable Energy Future." *Nature*, *488* (7411) (August 16), 294–303. doi:10.1038/nature11475.

Cossent, R., Olmos, L., Gómez, T., Mateo, C. & Frías, P. (2011). "Distribution Network Costs Under Different Penetration Levels of Distributed Generation." *European*

Transactions on Electrical Power, *21* (6) (September 1), 1869–1888. doi:10.1002/etep.503.

CPUC (California Public Utilities Commission). (2011a). *Attachment 2: Standardized Planning Assumptions (Part 2 – Renewables) for System Resource Plans.* R.10-05-006. San Francisco, CA: California Public Utilities Commission. http://docs.cpuc.ca.gov/efile/RULINGS/130670.pdf.

CPUC (California Public Utilities Commission). (2011b). *Qualifying Capacity Methodology Manual: Appendix B*. R.09-10-032. San Francisco, CA: California Public Utilities Commission. http://www.cpuc.ca.gov/NR/rdonlyres/2526B26C-BEEA-46FE-904F-A99D2F042FD8/0/AdoptedQCmethodologymanualfromD1006036APPENDIXB.doc.

Denholm, P. & Hand, M. (2011). "Grid Flexibility and Storage Required to Achieve Very High Penetration of Variable Renewable Electricity." *Energy Policy*, *39* (3) (March): 1817–1830. doi:16/j.enpol.2011.01.019.

Denholm, P. & Margolis, R. M. (2007). "Evaluating the Limits of Solar Photovoltaics (PV) in Electric Power Systems Utilizing Energy Storage and Other Enabling Technologies." *Energy Policy*, *35* (9) (September), 4424–4433. doi:10.1016/j.enpol.2007.03.004.

Denholm, P. & Mehos, M. (2011). *Enabling Greater Penetration of Solar Power via the Use of CSP with Thermal Energy Storage*. NREL/TP-6A20-52978. Golden, CO: National Renewable Energy Laboratory. www.nrel.gov/csp/pdfs/52978.pdf

Denholm, P., Margolis, R. M. & Milford, J. M. (2009). "Quantifying Avoided Fuel Use and Emissions from Solar Photovoltaic Generation in the Western United States." *Environmental Science & Technology*, *43* (1) (January 1): 226–232. doi:10.1021/es801216y.

DOE (U.S. Department of Energy). (2012). *SunShot Vision Study*. DOE/GO-102012-3037. Washington D.C.: U.S. Department of Energy. http://www1.eere.energy.gov/solar/sunshot/vision_study.html.

Doherty, R., Outhred, H. & O'Malley, M. (2006). "Establishing the Role That Wind Generation May Have in Future Generation Portfolios." *IEEE Transactions On Power Systems*, *21* (3), 1415–1422. doi:10.1109/TPWRS.2006.879258.

Dragoon, K. & Dvortsov, V. (2006). "Z-method for Power System Resource Adequacy Applications." *IEEE Transactions on Power Systems*, *21* (2), 982–988.

EASAC (European Academies Science Advisory Council). (2011). *Concentrating Solar Power: Its Potential Contribution to a Sustainable Energy Future*. EASAC Policy Report 16. Halle, Germany: EASAC. http://www.easac.eu/home/reports-and-statements/detail-view/article/concentratin.html.

EnerNex Corporation. (2009). *Solar Integration Study for Public Service Company of Colorado*. Denver, CO: Xcel Energy.

Fripp, M. (2012). "Switch: A Planning Tool for Power Systems with Large Shares of Intermittent Renewable Energy." *Environmental Science & Technology*, *46* (11), (June 5), 6371–6378. doi:10.1021/es204645c.

GE Energy. (2010). *Western Wind and Solar Integration Study*. NREL/SR-550-47434. Golden, CO: National Renewable Energy Laboratory. www.nrel.gov/wind/systemsintegration/wwsis.html.

Goodrich, A., James, T. & Woodhouse, M. (2012). *Residential, Commercial and Utility-Scale Photovoltaic (PV) System Prices in the United States: Current Drivers and Cost-*

Reduction Opportunities. NREL/TP-6A20-53347. Golden, CO: National Renewable Energy Laboratory. www.nrel.gov/docs/fy12osti/53347.pdf.

Gross, R., Heptonstall, P., Anderson, D., Green, T., Leach, M. & Skea, J. (2006). *The Costs and Impacts of Intermittency: An Assessment of the Evidence on the Costs and Impacts of Intermittent Generation on the British Electricity Network*. Imperial College London: UK Energy Research Centre (UKERC). www.ukerc.ac.uk/Downloads/PDF/06/ 0604Intermittency/0604IntermittencyReport.pdf.

Gross, R., Heptonstall, P., Leach, M., Anderson, D., Green, T. & Skea, J. (2007). "Renewables and the Grid: Understanding Intermittency." *Proceedings of the Institution of Civil Engineers, 160* (ENI) (February), 31–41. doi:10.1680/ener.2007.160.1.31.

Hirst, E. & Goldman, C. (1991). "Creating the Future: Integrated Resource Planning for Electric Utilities." *Annual Review of Energy and the Environment, 16* (1), 91–121. doi:10.1146/annurev.eg.16.110191.000515.

Hoff, T. (1988). "Calculating Photovoltaics' Value: A Utility Perspective." *IEEE Transactions on Energy Conversion, 3* (September), 491–495. doi:10.1109/60.8056.

Hoffmann, M. G., Sadovsky, A., Kintner-Meyer, M. C. & DeSteese, J. G. (2010). *Analysis Tools for Sizing and Placement of Energy Storage in Grid Applications, a Literature Review*. PNNL- 19703. Richland, WA: Pacific Northwest National Laboratory. www.pnl.gov/main/publications/external/technical_reports/PNNL-19703.pdf.

Holttinen, H., Meibom, P., Orths, A., Lange, B., O'Malley, M., Tande, J. O., Estanqueiro, A., et al. (2011). "Impacts of Large Amounts of Wind Power on Design and Operation of Power Systems, Results of IEA Collaboration." *Wind Energy, 14* (2) (March 1), 179–192. doi:10.1002/we.410.

Ibanez, E. & Milligan, M. (2012). *A Probabilistic Approach to Quantifying the Contribution of Variable Generation and Transmission to System Reliability*. NREL/CP-5500-56219. Golden, CO: National Renewable Energy Laboratory. www.nrel.gov/docs/fy12osti/56219.pdf.

IEA (International Energy Agency). (2011). *Renewable Energy: Markets and Prospects by Technology*. Paris, France: International Energy Agency. www.iea.org/publications/freepublications/publication/Renew_Tech.pdf.

Jin, S., Ryan, S., Watson, J. P. & Woodruff, D. (2011). "Modeling and Solving a Large-scale Generation Expansion Planning Problem Under Uncertainty." *Energy Systems, 2* (3), 209–242. doi:10.1007/s12667-011-0042-9.

Jones, R. (2012). "Diversity Benefit of Solar and Wind with Increasing Market Penetration." *Intersolar North America*. July 12.

Joskow, P. L. (2011). "Comparing the Costs of Intermittent and Dispatchable Electricity Generating Technologies." *The American Economic Review, 101* (3), (May), 238–241. doi:10.1257/aer.100.3.238.

Katiraei, K. F. & Aguero, J. R. (2011). "Solar PV Integration Challenges." *IEEE Power and Energy Magazine, 9* (3), (June), 62–71. doi:10.1109/MPE.2011.940579.

KEMA Consulting. (2003). *Analysis and Selection of Analytical Tools to Assess National-Interest Transmission Bottlenecks*. Berkeley, CA: Lawrence Berkeley National Laboratory. http://certs.lbl.gov/pdf/kema-nitb.pdf.

Lamont, A. D. (2008). "Assessing the Long-term System Value of Intermittent Electric Generation Technologies." *Energy Economics, 30* (3), (May), 1208–1231. doi:10.1016/j.eneco.2007.02.007.

Larsen, P. & Wilkerson, J. (2012). "Review of Western U.S. Electric Utility Resource Planning Information". Presentation to Committee on Regional Electric Power Cooperation, State- Provincial Steering Committee and Resource Planners' Forum, October 3. www.westgov.org/wieb/meetings/crepcfall2012/briefing/present/Larsen-Wilkerson.pdf

Lin, J. (2012). "Potential Impact of Solar Energy Penetration on PJM Electricity Market." *IEEE Systems Journal*, *6* (2), (June), 205 –212. doi:10.1109/JSYST.2011.2162885.

Madaeni, S. H., Sioshansi, R. & Denholm , P. (2012b). "How Thermal Energy Storage Enhances the Economic Viability of Concentrating Solar Power." *Proceedings of the IEEE*, *100* (2), (February), 335–347. doi:10.1109/JPROC.2011.2144950.

Madaeni, S. H., Sioshansi, R. & Denholm, P. (2012a). "Estimating the Capacity Value of Concentrating Solar Power Plants: A Case Study of the Southwestern United States." *IEEE Transactions on Power Systems*, *27* (2) (May), 1116 –1124. doi:10.1109/TPWRS.2011.2179071.

Madaeni, S. H., Sioshansi, R. & Denholm, P. (2012c). "Comparing Capacity Value Estimation Techniques for Photovoltaic Solar Power." *IEEE Journal of Photovoltaics*, PP (99), 1 –9. doi:10.1109/JPHOTOV.2012.2217114.

Milligan, M., Ela, E., Hodge, B., Kirby, B., Lew, D., Clark, C., DeCesaro, J. & Lynn, K. (2011). "Integration of Variable Generation, Cost-Causation & Integration Costs." *The Electricity Journal*, *24* (9) (November), 51–63. doi:10.1016/j.tej.2011.10.011.

Mills, A. & Wiser, R. (2010). *Implications of Wide-Area Geographic Diversity for Short-Term Variability of Solar Power*. LBNL-3884E. Berkeley, CA: Lawrence Berkeley National Laboratory. http://eetd.lbl.gov/ea/emp/reports/lbnl-3884e.pdf.

Mills, A. & Wiser, R. (2012). *Changes in the Economic Value of Variable Generation with Increasing Penetration Levels: A Pilot Case Study of California*. LBNL-5445E. Berkeley, CA: Lawrence Berkeley National Laboratory. http://eetd.lbl.gov/ea/emp/reports/lbnl-5445e.pdf.

Mills, A., Phadke, A. & Wiser, R. (2011). "Exploration of Resource and Transmission Expansion Decisions in the Western Renewable Energy Zone Initiative." *Energy Policy*, *39* (3), (March), 1732–1745. doi:10.1016/j.enpol.2011.01.002.

Müsgens, F. & Neuhoff, K. (2006). *Modelling Dynamic Constraints in Electricity Markets and the Costs of Uncertain Wind Output*. Working Paper. EPRG. Cambridge, UK. http://www.dspace.cam.ac.uk/bitstream/1810/131648/1/eprg0514.pdf.

Nagl, S., Fürsch, M., Jägemann, C. & Bettzüge, M. O. (2011). *The Economic Value of Storage in Renewable Power Systems-the Case of Thermal Energy Storage in Concentrating Solar Plants*. EWI Working Papers. Energiewirtschaftliches Institut an der Universitaet zu Koeln. http://www.ewi.uni- koeln.de/fileadmin/user_upload/Publikationen/Working_Paper/EWI_WP_11- 08_The_economic_value_of_storage.pdf.

Navigant Consulting, Sandia National Laboratories, Pacific Northwest National Laboratories and NV Energy. (2011). *Large-scale Solar Integration Study*. Las Vegas, Nevada: NV Energy. www.navigant.com/insights/library/energy/large-scale_pv_integration_study/.

Neuhoff, K., Ehrenmann, A., Butler, L., Cust, J., Hoexter, H., Keats, K., Kreczko, A. & Sinden, G. (2008). "Space and Time: Wind in an Investment Planning Model." *Energy Economics*, *30* (4), (July), 1990–2008. doi:10.1016/j.eneco.2007.07.009.

NREL (National Renewable Energy Laboratory). (2012). *Renewable Electricity Futures Study*. NREL/TP-6A20-52409. Golden, CO: National Renewable Energy Laboratory. www.nrel.gov/analysis/re_futures/.

Olson, A. & Jones, R. (2012). "Chasing Grid Parity: Understanding the Dynamic Value of Renewable Energy." *The Electricity Journal, 25* (3) (April), 17–27. doi:10.1016/j.tej.2012.03.001.

Palmintier, B. & Webster, M. (2011). "Impact of unit commitment constraints on generation expansion planning with renewables." In *2011 IEEE Power and Energy Society General Meeting*, 1–7. IEEE. doi:10.1109/PES.2011.6038963.

Pelland, S. & Abboud, I. (2008). "Comparing Photovoltaic Capacity Value Metrics: A Case Study for the City of Toronto." *Progress in Photovoltaics: Research and Applications, 16* (8), 715–724. doi:10.1002/pip.864.

Perez, R., Taylor, M., Hoff, T. & Ross, J. P. (2008). "Reaching Consensus in the Definition of Photovoltaics Capacity Credit in the USA: A Practical Application of Satellite-Derived Solar Resource Data." *IEEE Journal of Selected Topics in Applied Earth Observations and Remote Sensing, 1* (1), 28–33. doi:10.1109/JSTARS.2008.2004362.

Perez, R., Zweibel, K. & Hoff, T. E. (2011). "Solar Power Generation in the US: Too Expensive, or a Bargain?" *Energy Policy, 39* (11), (November), 7290–7297. doi:10.1016/j.enpol.2011.08.052.

Rábago, K. R., Libby, L., Harvey, T., Norris, B. & Hoff, T. (2012). "Designing Austin Energy's Solar Tariff Using a Distributed PV Value Calculator." In *Proceedings of World Renewable Energy Forum 2012*. www.cleanpower.com/wp- content/uploads/090_DesigningAustinEnergysSolarTariff.pdf.

Rogers, J. & Porter, K. (2012). *Summary of Time Period-Based and Other Approximation Methods for Determining the Capacity Value of Wind and Solar in the United States: September 2010-February 2012.* NREL/SR-5500-54338. Golden, CO: National Renewable Energy Laboratory. www.nrel.gov/docs/fy12osti/54338.pdf.

Schwartz, L. (2012). *Renewable Resources and Transmission in the West: Interviews on the Western Renewable Energy Zones Initiative*. Western Governors' Association. http://www.westgov.org/reports.

SEPA (Solar Electric Power Association). (2009). *Utility Solar Procurement Study: Solar Electricity in the Utility Market*. Washington D.C.: Solar Electric Power Association. www.solarelectricpower.org/resources/publications.aspx - Utility_Solar_Procurement_Study_January2009.

Shortt, A., Kiviluoma, J. & O'Malley, M. (2012). "Accommodating Variability in Generation Planning." *IEEE Transactions on Power Systems*, PP(99), 1–12. doi:10.1109/TPWRS.2012.2202925.

Shugar, S. & Hoff, T. (1993). "Grid-support Photovoltaics: Evaluation of Criteria and Methods to Assess Empirically the Local and System Benefits to Electric Utilities." *Progress in Photovoltaics: Research and Applications, 1* (3), 233–250. doi:10.1002/pip.4670010308.

Sioshansi, R. & Denholm, P. (2010). "The Value of Concentrating Solar Power and Thermal Energy Storage." *IEEE Transactions on Sustainable Energy, 1* (3) (October), 173 –183. doi:10.1109/TSTE.2010.2052078.

Sioshansi, R., Denholm, P. & Jenkin, T. (2012). "Market and Policy Barriers to Deployment of Energy Storage." *Economics of Energy & Environmental Policy, 1* (2), (April). doi: 10.5547/2160- 5890.1.2.4

Söder, L. (2005). "The Value of Wind Power." In *Wind Power in Power Systems*, ed. Thomas Ackermann Msc Researcher, 169–195. John Wiley & Sons, Ltd. doi:10.1002/0470012684.ch9.

Staffell, I. & Green, R. (2012). *Is There Still Merit in the Merit Order Stack? The Impact of Dynamic Constraints on Optimal Plant Mix*. United Kingdom: British Institute of Energy Economics. www.biee.org/wpcms/wp-content/uploads/Staffell-Merit-in-the-Merit-Order-Stack.pdf.

Stoft, S. (2002). *Power System Economics: Designing Markets for Electricity*. Wiley/IEEE.

Vithayasrichareon, P. & MacGill, I. F. (2012). "A Monte Carlo Based Decision-support Tool for Assessing Generation Portfolios in Future Carbon Constrained Electricity Industries." *Energy Policy, 41* (February), 374–392. doi:10.1016/j.enpol.2011.10.060.

Wiser, R. & Barbose, G. (2008). *Renewable Portfolio Standards in the United States - A Status Report with Data Through 2007*. LBNL-154E. Berkeley, CA: Lawrence Berkeley National Laboratory. http://eetd.lbl.gov/ea/ems/reports/lbnl-154e.pdf.

Wiser, R. & Bolinger, M. (2006). "Balancing Cost and Risk: The Treatment of Renewable Energy in Western Utility Resource Plans." *The Electricity Journal, 19* (1), 48–59.

Wiser, R., Barbose, G. & Holt, E. (2011). "Supporting Solar Power in Renewables Portfolio Standards: Experience from the United States." *Energy Policy, 39* (7) (July), 3894–3905. doi:10.1016/j.enpol.2010.11.025.

Xcel Energy. (2011). *Wind Induced Coal Plant Cycling Costs and the Implications of Wind Curtailment for Public Service of Colorado*. Denver, CO: Xcel Energy. www.uwig.org/11M- 710E_WindInducedCoalPlantCycling.pdf.

End Notes

[1] Natural gas augmentation is a modification to CSP plants in which the boiler used in the steam cycle can burn natural gas. This allows the CSP plant to operate when insolation is low. The low efficiency of using natural gas in a steam boiler compared with a combined cycle natural gas plant typically means that natural gas augmentation is only used during times when insolation is low and more efficient power plants are already fully deployed.

[2] Madaeni et al. (2012b) only present changes in the value of CSP with thermal storage and the solar multiplier in $/kW terms, not $/MWh terms. Based on simple calculations using their figures it appears that adding 4 hours of thermal storage and increasing the solar multiple from 1.5 to 2.0 increases the value of CSP by about $10/MWh in Nevada and by about $20/MWh in Death Valley (about half of that increase in the value with thermal storage comes from the sale of ancillary services based on the CAISO ancillary service prices between 2001 and 2005).

[3] Austin Energy is beginning to offer a solar tariff to PV customers that compensates PV generation at a rate set to the current year's estimated value of solar to the LSE (Rábago et al. 2012). We do not include this tariff in our analysis because it focuses primarily on designing a fair compensation scheme rather than planning a portfolio of resources. Many concepts used in the design of the Austin Energy tariff are similar to the concepts reviewed in this report.

[4] Of course, solar is but one of many considerations in LSE planning studies; the desire for better representation of solar should be balanced with practical constraints due to limitations in available tools, methods, and time.

[5] "Capacity value" here refers to the economic value related to the savings associated with solar displacing the need to procure other sources of capacity (with units of $/MWh). "Load-carrying capability" represents the amount of additional load that can be reliably met when solar is added to a portfolio (with units of MW). "Capacity credit" refers to the LSE's assumed/estimated load-carrying capability of solar per unit of solar nameplate capacity (represented as a % of nameplate capacity). We maintain these definitions consistently through this

document but note that many LSEs and researchers use these terms differently. Most notably, the term we refer to here as "capacity credit" is synonymous with what others sometimes call "capacity value." When we use "capacity value" we mean the economic savings from avoiding other capacity resources, similar to how we use "energy value" to describe the economic savings from reducing production costs.

[6] Joskow (2011) states, for example, "The most widely used metric for comparing the 'competitiveness' of different generating technologies is the estimated 'levelized cost' per megawatt-hour (MWh) supplied....competitive procurement programs run by utilities to meet renewable electricity purchase mandates often use auction mechanisms that effectively choose the supply offers with the lowest levelized cost per MWh regardless of when it is supplied."

[7] We only reviewed the southern Nevada plan for NV Energy in this report.

[8] Developers may construct a plant that will then be owned by the LSE, construct a plant whose output will be sold under long-term contract to the LSE, or contract to sell power to the LSE from an existing facility on a short-term (a few years or less) or long-term (more than 10 years) basis. In some cases the LSE may build the plant itself if the options from project developers are not more attractive.

[9] Candidate portfolios are groups of demand-side and supply-side resource options, including solar, that are evaluated in the planning studies and used to justify the LSEs preferred portfolio or strategy going forward.

[10] Generating resources did not necessarily need to be under long-term contract to the LSE to count toward satisfying its peak demand needs, but studies usually identified the total capacity that needed to be at least under short-term contract.

[11] One exception was the Northwest Power and Conservation Council (NPCC), which is primarily concerned with adequate energy rather than adequate capacity. The Pacific Northwest has significant amounts of energy-limited hydropower that can provide large amounts of power but not for an extended period. The "feasible" portfolios, therefore, are not required to satisfy particular capacity needs; instead, portfolios with too little generation result in large exposure to market purchases that may be uneconomic owing to the assumed high volatility of market prices.

[12] A number of LSEs indicated their preference for the use of sensitivity cases using bookend scenarios over more complex Monte-Carlo simulations. Their reason was the transparency and clarity that single-variable sensitivity cases provided: the analysts could understand the degree to which the PVRR would change under the given sensitivity conditions. For example PGE stated:

While we believe that both stochastic and deterministic scenario analyses provide important insights for assessing the performance and reliability of a portfolio over time, we have found that the most substantial risks in connection with making future resource choices are those associated with large fundamental or structural shifts – the types of risk best described through scenario analysis. As a result, we believe that scenario analysis should be given the primary emphasis in our overall portfolio risk evaluation. However, we do also continue to consider the instructive value from the stochastic analysis.

Ultimately no degree of modeling and analysis can account for all possible future uncertainties. Modeling by its nature only provides an estimate or range of estimates of future results. Nevertheless, we believe that a well- reasoned and complementary application of both scenario and stochastic analysis can provide useful insights about how a candidate portfolio is likely to perform in the future.

[13] Common risk measures included the value at risk (VaR), which is the PVRR at some particular high percentile of Monte-Carlo draws (e.g., the PVRR for the 90th percentile), and the conditional value at risk (CVaR), which is the average PVRR for the most costly fraction of all draws (e.g., the average of the PVRR for the most costly 10% of draws).

[14] PGE, for example, developed a very detailed method for identifying its preferred portfolio from the many that were evaluated. Both scenario analysis based on several key uncertainties (future fuel prices, carbon prices and carbon price timing, availability of renewable tax incentives, and wholesale market prices) and Monte-Carlo analysis were used to evaluate 15 portfolios created by PGE. PGE then applied a scoring to each portfolio based on its ranking in terms of cost, risk, reliability, and resource diversity. The preferred portfolio was the one with the highest score by weighting each of the different categories as follows: 20% based on reliability and diversity metrics, 30% based on portfolio risk metrics, and 50% based on expected cost.

[15] The levelized fixed cost could be replaced with the amount the LSE would pay to independent PV owners if the LSE procured the PV power through long-term PPAs instead of building and owning the PV.

[16] Recent analysis of high renewable scenarios in the United States can help identify regions suitable for CSP. The DOE *SunShot Vision Study* included CSP in California, Arizona, Nevada, Utah, Colorado, New Mexico, and Texas in a 2050 scenario (DOE 2012). The NREL *Renewable Electricity Futures Study* identified a similar geographic distribution of CSP plants in a 2050 scenario where 80% of U.S. electricity is provided by renewable resources (NREL 2012). The same studies included PV in every state in the scenarios.

[17] One exception was the NPCC model, used in a region with large amounts of energy-limited hydro, where a portfolio with low amounts of new generation capacity was found to expose the loads to high prices in the wholesale market. To avoid exposure to these high prices, the model consistently chose portfolios with generating capacity that exceeded minimum levels needed for reliability purposes (i.e., the capacity constraint was not binding).

[18] A 0% capacity credit was assigned to small PV in the CA IOU process, which assumed that small PV sited on the distribution system could not be counted toward meeting the peak load and planning reserve margin since the resources can connect to the distribution system through a process that does not include a deliverability study required under the California Resource Adequacy program (CPUC 2011a). It is expected that the CAISO will conduct deliverability studies in the near future, however, and even small PV will become eligible for a positive capacity credit.

[19] The capacity credits reported for the CA IOUs in this section were extracted from the *33% RPS Calculator* used in the 2010 Long-Term Procurement Planning process. Once solar (and wind) facilities are operating, the capacity credit applied to the resource for the purposes of the California Resource Adequacy process will be based on the actual operating history of the plant during peak-load periods (CPUC 2011b).

[20] Although PGE applied the 5% capacity credit to PV in its planning studies, it did not use a quantitative evaluation method to support this value. Instead the value was based on PGE's judgment that, due to the winter peaking nature of PGE's demand, the capacity credit for PV was expected to be similar to the capacity credit for wind (which was previously estimated to be 5% in this region).

[21] It is not clear if the capacity credit of CSP without thermal storage or natural gas augmentation should be higher than the capacity credit for tracking PV. Comparison of the capacity credit of single-axis tracking PV (70%-87% in Madaeni et al. 2012c) to the capacity credit of CSP with a solar multiplier of 1.5 (53%-75% in Madaeni et al. 2012a) for various solar sites in WECC suggests the opposite: Madaeni et al. show higher capacity credits for single-axis PV than for CSP without thermal storage. Reasons that could justify a higher capacity credit for CSP without thermal storage relative to single-axis PV in the same region include:

- A CSP plant with a solar multiple (the ratio of the thermal energy from the solar field under standard conditions to the thermal energy required to operate the powerblock at its rated capacity) of greater than 1.0 would have more generation in the evening: Madaeni et al. (2012a) find that for CSP sites in the Southwest increasing the solar multiple from 1.0 to 1.5 can increase the capacity credit from around 60% to around 80%. Increasing the solar multiple of a CSP plant leads to unused solar heat during high insolation periods but increased production in the early evening and morning relative to a CSP plant with a lower solar multiple. The generation profile of single-axis tracking PV is likely to match the generation profile of CSP with a solar multiple closer to 1.0.
- The thermal inertia of a CSP plant would allow it to continue to produce energy for roughly 15 minutes with passing clouds, whereas PV plant output would drop off much more quickly.

On the other hand, single-axis PV could have a higher capacity credit than CSP without thermal storage for the following reasons:

- CSP only uses direct-normal insolation, whereas PV uses direct and diffuse, allowing PV to generate power even from scattered light.
- CSP plants have a minimum generation level at which insufficient thermal input requires the CSP plant to stop producing power. This could lead to a CSP plant producing zero power and a single-axis tracking plant producing some power just before sunset.

Given these conflicting trends in estimates of capacity credits, any distinct difference between the capacity credit of tracking PV and the capacity credit of CSP without thermal storage or natural gas augmentation used by LSEs in planning studies should be confirmed and justified with further detailed studies.

[22] On the other hand, if both have high PV penetrations, the incremental capacity credit for more PV will be lower for either LSE, lessening any potential benefits from trade.

[23] The wholesale prices generated by the model, however, are not sufficient to cover the fixed cost of all new investments. PGE states: "Given these assumptions, the AURORAxmp forecasted electricity price is generally not adequate to achieve a positive return of and return on invested capital for new resources. Therefore, it is assumed that costs, particularly for capacity, would need to be recovered through regulation or a separate capacity market."

[24] Although some LSEs directly increased AS requirements in their production cost model evaluation of different portfolios, none directly evaluated the cost of day-ahead or multiple hour-ahead forecast errors in the production cost model runs. Numerous integration studies that have focused on the technical feasibility of high renewable penetration levels or on estimating integration cost adjustments (as described in Section 4.5.2) have included both AS requirements and the impact of day-ahead forecast errors on production costs.

[25] One LSE planning study, the CA IOU process, also allowed for the increase in operating reserves and variability due to solar (and wind) to potentially increase the need for additional CTs. The degree to which additional CTs were needed due to integration-related needs versus a declining capacity contribution of solar with increasing penetration is still being resolved at the time of the study publication. Additional analysis in the next planning cycle is expected to resolve these ambiguities.

[26] It makes sense that LSEs did not attempt to isolate the contribution of particular resources to reducing the overall risk of candidate portfolios since risk is primarily defined in terms of overall portfolios. It would be possible to estimate the risk-reduction benefit of a particular technology, however, by comparing the risk of a portfolio

with and without the technology. Estimates of the risk-reduction benefit of particular technologies could then be used to identify which resources to include in portfolios (Step 2) and/or to compare resources during procurement (Step 5).

[27] The methodology used by the California Public Utilities Commission to calculate the qualifying capacity for solar and wind projects contracted or owned by the IOUs (the resource adequacy or RA program) does include a diversity benefit. This approach is used in accounting for the qualifying capacity of projects once built; it does not appear to be used in the planning study. In the RA program, the diversity benefit is defined as the difference between the qualifying capacity estimated for all wind and solar resources combined less the sum of the qualifying capacity estimated for each wind or solar project individually. This diversity benefit is allocated to each wind and solar project on a per-energy-unit basis (CPUC 2011b).

[28] Regional Portfolio Model is an in-house, Excel-based capacity-expansion model.

[29] Capacity Expansion is a precursor to the System Optimizer model.

[30] The full set of inter-relationships between different proceedings within the CPUC and various other state and federal agencies involved with the California IOU planning and procurement process is much more involved than described in this limited review focused on solar. For a more detailed overview of the CA process see the supporting documents reviewed in this report and a recent overview prepared for the Regulatory Assistance Project (Burgess et al. 2012).

[31] Changes to the NQC methodology to reflect changes in capacity credit with increasing penetration would need to occur through the Resource Adequacy program.

[32] In addition to the detail provided in the 2012 APS IRP, APS resource planning staff provided supplementary information during the review process that is reflected in this summary.

INDEX

D

E

F

O

P

Q

R